Australian Army Operational Shoulder Patches
Post 2000

Second Edition, June 2025

Butler D. Brisbane 2025

Copyright Notice: © 2025

The Author asserts the moral and other rights to be identified as the creator and originator of this work, pursuant to the Copyright Act 1968 (C'th) and in accordance with the Berne Convention for the Protection of Literary and Artistic Works. The Bern Convention is administered by The World Intellectual Property Organization (WIPO), a specialized agency of the United Nations, which is dedicated to developing a balanced and accessible international intellectual property (IP) system, which rewards creativity, stimulates innovation and contributes to economic development while safeguarding the public interest.

All rights in this work are reserved to the Author. No part of this publication may be reproduced, stored in a retrieval system, or be transmitted in any form or by any means, without the prior written permission of the Copyright holder, nor may this publication be circulated in any form of binding or cover other than that in which it was originally published and without a similar condition including this condition, being imposed upon the subsequent publisher.

Enquiries with respect to property rights in this work may be directed to the author at:

- cavdb_71@hotmail.com
- info@bpcmilitaria.com.au

Published by the Author, Brisbane Australia, 2025

- http://www.bpcmilitaria.com.au
- http://www.colourpatch.com.au

Design by Iain Anderson

- http://funwithstuff.com

National Library of Australia Cataloguing-in-Publication entry

Creator:	Butler, David Anthony, author.
Title:	Australian Army Operational Shoulder Patches : (Post 2000) / David Anthony Butler (author).
ISBN:	9780975637616 (paperback, 2nd edition)
Notes:	Includes bibliographical references.
Subjects:	Australia. Army--Insignia--History. Australia. Army--Medals, badges, decorations, etc.--History. Australia. Army--Insignia--Pictorial works. Australia. Army--Medals, badges, decorations, etc.--Pictorial works. Armies--Insignia.
Other Creators/Contributors:	NIL
Published:	Browns Plains, QLD : CharlieBravoBooks.com, 2025.
Summary:	The purpose of the book is to document the Australian Army's Official and Unofficial Operational Shoulder Patches (OSPs). This publication will allow Users to correctly identify Operational patches by Theatre, Deployment and, to a degree, in the order they were issued. This publication is not a definitive list of Operational patches but a consolidation of the items available at the time this work was compiled.

A catalogue record for this book is available from the National Library of Australia

Introduction

The purpose of the book is to document the Australian Army's Official and Unofficial Operational Shoulder Patches (OSPs). This publication will allow Users to correctly identify Operational patches by Theatre, Deployment and, to a degree, in the order they were issued. This publication is not a definitive list of Operational patches but a consolidation of the items available at the time this work was compiled.

This work has a start date of 2000 which is when the Australian Army was deploying in large numbers after Interfet in 1999. I have titled the book "Australian Army Operational Shoulder Patches" even though I have included some Air Force and Navy patches, however 90–95% of the patches are Army oriented. If this was to change in a future revised edition, I would then change the name to ADF. Earlier deployments had a different design each rotation, to save money the ADF then decided to use generic deployment patches ie., Reconstruction Task Force had rotation patches for the 4 rotations and then a generic patch was worn for the 2 rotations of Mentoring Reconstruction Task Force and then the 5 rotations of the Mentoring Task Force.

Early deployment patches were of different sizes and shapes, but slowly over time they have conformed to the ADF regulation sizes. The Army patch is 55mm x 75mm and is khaki bordered, the RAN patch is 50mm x 100mm with a yellow border and the RAAF patch is 50mm x 80mm with a black border.

I have put a question mark on the patches that I deem "suss"... I am only going from my own information.

If you have any patches not included, or want to send feedback please send to info@bpcmilitaria.com.au.

The following is an extract from the Army Dress Manual, Chapter 3 Para 3.165 to 3.167:

Shoulder Patches

Operational shoulder patch

3.165 Only an authorised Operational Shoulder Patch (OSP) may be worn. The OSP is worn with Field Dress (AMCU) whilst deployed on operations outside of Australia. The OSP is not to be worn in barracks. Only one patch is to be worn on the right sleeve of the AMCU shirt. The OSP is not to be sewn onto the sleeves of the AMCU shirt.

3.166 OSPs for Force Elements (FE) deploying are requested through the Mounting HQ and approved by HQ Joint Operations Command (JOC) in the same manner as a USP. OSP for FE deploying on operations are funded from operational sustainment funding and procured for either an FE, eg MTF or an entire operation, Op ANODE or Op ASTUTE. OSP that contain the following will not be approved:

a. rotation numbers

b. unit specific references, mascots, etc.

3.167 Approved OSP are to be procured, catalogued, and added to Block Scale 3004-08. This will ensure that the patch is procured in sufficient quantity to kit multiple rotations with additional held for sustainment stock. Maintenance of the OSP once approved is a unit responsibility. The manufacturers will need to be provided with the design specification by the unit. Once approved and manufactured, units are to provide one OSP with coloured design and authorised colour codes to DGAPC for quality control and a one OSP to the Army History Unit (AHU) for central historical collection.

https://www.army.gov.au/sites/default/files/2023-08/Army-Dress-Manual-AL5.pdf

Biographic Notes

My name is David Butler and I was born in RAF Hospital in Changi Singapore in 1971. My father (Arthur Butler) was a member of 6 RAR at that time, when the Battalion was posted to Singapore for 2 ½ years.

I grew up as an 'Army brat', until I joined the Army Reserve 2/14 LH (QMI) as an RAAC Crewman at the age of 17 and after 3 ½ years I joined the Australian Regular Army (ARA) in 1992. After Kapooka and initial employment training (IETs) for RAAC, my first posting was to B Sqn 3rd/4th Cav Regt... then 2nd Cavalry Regt, before I Corps transferred to Royal Australian Electrical and Mechanical Engineers (RAEME) to become a Vehicle Mechanic. Once qualified as a Vehicle Mechanic, I was posted back to B Sqn 3rd/4th Cavalry Regt, where I picked up my first deployment to East Timor (AUSBAT 8) after 11 yrs in the ARA.

In 2006, I was posted to 2/14 LH (QMI) where I deployed on AMTG 3/OBG(W)-1 and OBG(W)-4 to Iraq. In 2008 I was posted to 3rd/4th Cavalry Regt for the third time and deployed on MRTF-2 and MTF 3 to Afghanistan. I discharged from the ARA in October 2012 and after doing some Army Reserve service, my final day in uniform was in May 2017 at 2/14 LH(QMI)... back where I had started. The best decision I ever made was to Corps Transfer to RAEME and to get a trade, as that decision got me my deployments and my current job of a Diesel Fitter in the mining industry. During this time, I survived several ugly divorces and yet, I raised a bunch of beautiful children, who survived my Army career and made me a doting father at the same time.

So, we come to the matter of collecting various types of Australian Army insignia. I first started collecting Militaria in 1993 when my father gave me some ARVN badges which he had collected (whilst he was twice deployed to Vietnam), along with some of his accumulated Australian Badges. In short order, I started collecting the (then) current Anodised Aluminium insignia of the Australian Army, until I had a near complete set of hat badges, collars and metal titles. My collecting interests continued to focus on metal insignia, until I deployed to Iraq in 2006. At that time, I started collecting the patches Australians were wearing in theatre. Later, my collecting field has grown to Australian Military Badges Pre-Federation through to the current day... including: Australian Defence Force Patches (Official, Operational and Unofficial), Military Buttons and Australian Military Challenge Coins. In 2012 I bought the 'colourpatch' website from Phil Blackwell and from then I started an Excel spreadsheet checklist for USPs. With steady help from Phil Blackwell, Mike Gordan and Andrew Lam, I have been able to expand that checklist into this book. A special thanks to Nick Fletcher and Danielle Cassar for allowing me to access to the Australian War Memorial collection of patches. In 2013 my 'bpcmilitaria' (**B**adges, **P**atches and **C**oins) website was created to sell all the other Miitaria not listed on the 'colourpatch' website.

Ultimately, I hope this book will be of some help to you, just as it has helped me to structure my own collection.

Future work will be on Australian Army Badges 1901/03–1952.

Previous Work:

- *Australian Army Unit and Tri Service Shoulder Patches: Post 2010*
- *Australian Army Operational Shoulder Patches: Post 2000*
- *Challenge Coins of the Australian Defence Force*
- *Australian Army Unit and tri-Service Shoulder Patches Post 2010 (Official and Unofficial)*

Happy Collecting.
David (Butts) Butler.

Table of Contents

Introduction	3
Biographic Notes	4

Afghanistan 6
- Joint Task Force Paladin 6
- Special Operations Task Group 6
- Headquarters 18
- Combined Team Uruzgan 24
- Force Support Unit 25
- Rotary Wing Group 25
- Artillery 28
- Coalition Advisory Team 29
- Medical 31
- UAV 32
- RTF MRTF MTF 34
- Multi National Base Command — TK 44
- National Emergency Operation 45
- Misc 45

Australian Operations 51
- JTF 114 Op Gold Sydney Olympics 2000 51
- JTF 629 COVID-19 Assist 51
- JTF 634 Op Deluge 52
- 635 Op Acolyte Melb C'Wealth Games 2006 53
- JTF 637 Op Atlas Gold Coast C'Wealth Games 2018 53
- JTF 637 Op Queenslander 54
- JTF 639 Op Resolute Border Protection 54
- JTF 640 Op Norwich 57
- JTF 641 Op Outreach 58
- JTF 644 Op Amulet 58
- JTF 646 Op Invincible Invictus Games 2018 59
- JTF 662 Operation Vic Fire Assist 2009 59
- JTF 664 60
- JTF 664 Op Cyclone Assist 2017 (TC Debbie) 60
- JTF 665 Op Parapet 61
- JTF 665 Op Testament 61
- Op Bushfire Assist 2019-20 62

Indo-Pacific Region 63
- Bougainville 63
- Indo Pacific Endeavour 64
- JTF 629 Op Samoa Assist 2009 64
- JTF 630 Op Pacific Assist 65
- JTF 637 Pacific Islands Support 65
- JTF 658 MH370 International Search Operation Southern Indian Ocean .. 66
- Malaysia 67
- Papau New Guinea 70
- Phillippines 72
- Solomons 73
- Timor Leste 74
- Tonga 79
- TU 659 Op Argo 79
- TU 659 Op Argo 80

IRAQ 81
- Operation Bastille 81
- Operation Falconer 81
- Operation Catalyst — Headquarters 82
- Operation Catalyst — Engineers 85
- Operation Catalyst — AATTI 87
- Operation Catalyst — SECDET 89
- Operation Catalyst — AMTG OBG FET 91
- Operation Catalyst — Novelty 102
- Operation Inherent Resolve — Headquarters 104
- Operation Inherent Resolve — TAJI 105
- Operation Inherent Resolve — Novelty 114
- Operation Operation Okra — SOTG 114
- Headquarters 116
- Engineers 119
- FLLA FSU FSE 120
- FCU TCU 126
- Misc 129
- ANF 134
- Novelty 138
- Special Forces 141
- RAAF 141
- RAN 149

Other Operations 152
- Nepal 152
- Pakistan 152
- Sinai 153
- Ukraine Support 154

United Nations 155

Glossary 158

AFGHANISTAN

Joint Task Force Paladin

Joint Task Force Paladin

AFGHANISTAN

Special Operations Task Group

TF 637 Special Operations Task Group
Round Coloured V1

TF 637 Special Operations Task Group
Round Coloured V2

TF 637 Special Operations Task Group
Round Subdued V1

TF 637 Special Operations Task Group
Round Subdued V2

TF 637 Special Operations Task Group
Round Subdued V3

SOTG — Afghan Rot 1 V1

SOTG — Afghan Rot 1 V2

TF 66 SOTG Coloured V1

TF 66 SOTG Coloured V2

TF 66 SOTG Coloured V3

TF 66 SOTG Coloured V4

TF 66 SOTG Coloured V5

TF 66 SOTG Coloured V6

TF 66 SOTG Subdued V1

TF 66 SOTG Subdued V2

TF 66 SOTG Subdued V3

TF 66 SOTG Subdued V4

TF 66 SOTG Subdued V5

TF 66 SOTG Subdued V6

TF 66 SOTG Subdued V7

TF 66 SOTG Subdued V8

Afghanistan

SOTG — Afghan 1 Sqn SASR

SOTG — Afghan 2 Sqn SASR V1

SOTG — Afghan 2 Sqn SASR V2

SOTG — Afghan E Tp 2 SAS SQN

SOTG — Afghan 3 Sqn SASR

SOTG — Afghan 3 Sqn SASR FAKE V1

SOTG — Afghan 3 Sqn SASR FAKE V2

SOTG — Afghan 3 Sqn SASR FAKE V3

SOTG — Afghan 4 RAR RAEME DPDU

SOTG — Afghan 4 RAR RAEME DPCU

SOTG — Afghan 1 Cdo Coy 1 Cdo Regt

SOTG — Afghan 2 Cdo Coy 1 Cdo Regt

SOTG — Afghan 1 Commando Regt

SOTG — Afghan 1st Commando Regt

SOTG — Afghan A Coy 2 Cdo Shield V1

Afghanistan

SOTG — Afghan A Coy 2 Cdo Shield V2

SOTG — Afghan A Coy 2 Cdo Shield V3

SOTG — Afghan A Coy 2 Cdo Shield V4

SOTG — Afghan A Coy 2 Cdo Shield V5

SOTG — Afghan A Coy 2 Cdo Shield V6

SOTG — Afghan A Coy 2 Cdo Rect

SOTG — Afghan A Coy 2 Cdo ANF V1

SOTG — Afghan Oscar Pl A Coy 2 Cdo

SOTG — Afghan 2 Cdo Alpha

SOTG — Afghan B Coy 2 Cdo

SOTG — Afghan B Coy 2 Cdo

SOTG — Afghan C Coy 2 Cdo Round Large V1

SOTG — Afghan C Coy 2 Cdo Round Large V2

SOTG — Afghan C Coy 2 Cdo Round Small V2

SOTG — Afghan C Coy 2 Cdo Round Small

Afghanistan

SOTG — Afghan T Pl C Coy 2 Cdo

SOTG — Afghan C Coy 2 Cdo Rect V1

SOTG — Afghan C Coy 2 Cdo Rect V1

SOTG — Afghan C Coy 2 Cdo Regt V2

SOTG — Afghan D Coy 2 Cdo V1

SOTG — Afghan D Coy 2 Cdo V2

SOTG — Afghan D Coy 2 Cdo V3

SOTG — Afghan D Coy 2 Cdo Shield

SOTG — Afghan A Coy 2 Cdo Skull

SOTG — Afghan 2 Cdo Foxtrot

SOTG — Afghan 2 Cdo Callsign F 14

SOTG — Afghan 2 Cdo Callsign F 16

SOTG — Afghan Rot Elephant Skull

SOTG — Afghan Q Store 2 Cdo Regt Round

SOTG — Afghan Q Store 2 Cdo Regt Rect

Afghanistan

SOTG — Afghan 2 Cdo Double Diamond DPDU

SOTG — Afghan 2 Cdo Double Diamond DPCU

SOTG — Afghan Incident Response Regiment

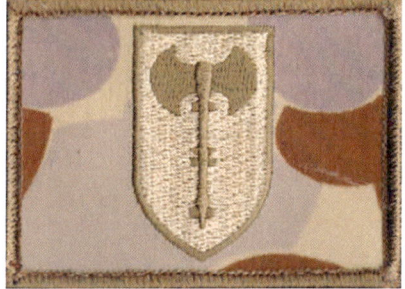
SOTG — Afghan SOER DPDU

SOTG — Afghan A Coy SOER MC V1

SOTG — Afghan A Coy SOER MC V2

SOTG — Afghan Rot 2 RAEME WKSPS

SOTG — Afghan Rot 2 SIGS V1

SOTG — Afghan Rot 2 SIGS V2

SOTG — Afghan Rot 6 SIGS V1

SOTG — Afghan Rot 6 SIGS V2

SOTG — Afghan Rot 8 RAEME Wksps

SOTG — Afghan Rot 9 RAEME Wksps

SOTG — Afghan Rot 10 RAEME Wksps

SOTG — Rot 15/16 Bushmaster Crews 3/4 Cav V1

Afghanistan

SOTG — Rot 15/16 Bushmaster Crews 3/4 Cav V2

SOTG — Rot 15/16 ISAF SOF Shield

SOTG — Afghan Rot 16 TASK V1

SOTG — Afghan Rot 16 TSAK V1

SOTG — Afghan Rot 16 TSAK V2

SOTG — Afghan Rot 16 TSAK V3

SOTG — Afghan Rot 18

SOTG — Afghan Rot 18 D Coy 2 Cdo

SOTG — Afghan Rot 18 6 Tp A Coy SOER V1

SOTG — Afghan Rot 18 6 Tp A Coy SOER V2

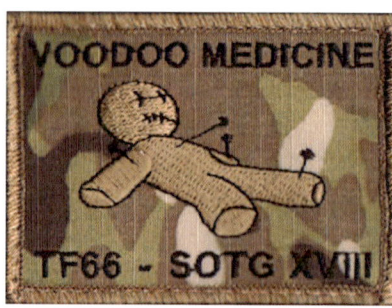
SOTG — Afghan Rot 18 Voodoo Medicine

SOTG — Afghan Rot 20 Voodoo Medicine

SOTG — Afghan JTAC

SOTG — Afghan TUAV

SOTG — RECON Aust-Afghan Flag

Afghanistan

SOTG — RECON Afghan Flag

SOTG — Bushmaster Crews 6 RAR

SOTG — Afghan Taxi Service

SOTG — Afghan Team Ramrod

SOTG — Afghan Rule of Law Coloured

SOTG — Afghan Rule of Law Subdued V1

SOTG — Afghan Rule of Law Subdued V2

TF 31 SOTG Rule of Law

TF 66 SOTG Force Insert/Extract Group V1

TF 66 SOTG Force Insert/Extract Group V2

SOTG — Afghan Voodoo Medicine Coloured

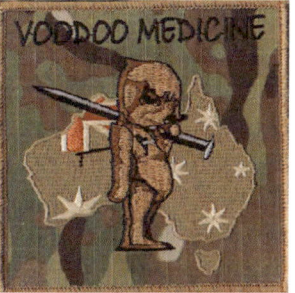
SOTG — Afghan Voodoo Medicine Subdued

SOTG — Afghan Voodoo Doll Round V1

SOTG — Afghan Voodoo Doll Round V2

SOTG — Afghan Voodoo Doll Round V3

Afghanistan

SOTG — Afghan Rot 3 Air Force Imagery Analyst (DPCU) Var 1

SOTG — Afghan Rot 3 Air Force Imagery Analyst (DPCU) Var 2

SOTG — Afghan Rot 3 Air Force Imagery Analyst (DPDU) Var 1

SOTG — Afghan Rot 3 Air Force Imagery Analyst (DPDU) Var 2

SOTG — Afghan Rot 3 Air Force Imagery Analyst (Grey)

SOTG — Afghan Rot 3 Air Force Imagery Analyst (Black)

SOTG — Afghan Rot 3 Air Force Imagery Analyst (Multicam)

SOTG — Afghan Rot 3 Air Force Imagery Analyst

SOTG Aust/Afghan Flags

SOTG Aust/Afghan Flag V1

SOTG Aust/Afghan Flag V2

SOTG Aust/Afghan Flag V3

SOTG Aust/Afghan Flag V4

SOTG Aust/Afghan Flag V5

SOTG Aust/Afghan Flag V6

SOTG Aust/Afghan Flag V7

SOTG — Afghan A Coy 2 Cdo Helicopters 1

SOTG — Afghan A Coy 2 Cdo Helicopters 2

SOTG — Afghan US Chinook Airlift V1

SOTG — Afghan US Chinook Airlift V2

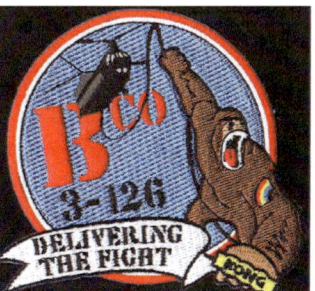
SOTG — Afghan US Chinook Airlift V3

SOTG — Afghan US Chinook Airlift V4

SOTG — Afghan US Chinook Airlift V5

SOTG — Afghan US Chinook Airlift V6

SOTG — Afghan US Chinook Airlift V7

SOTG — Afghan US Chinook Airlift V8

SOTG — Afghan US Chinook Airlift V9

SOTG — Afghan US Chinook Airlift V10

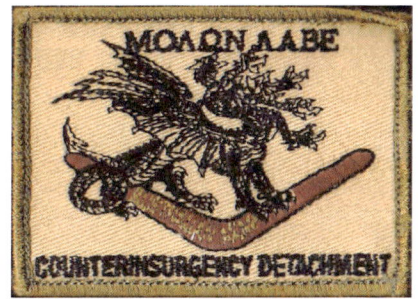
SOTG Counterinsurgency Det

Afghan Special Response Team Round V1

Afghanistan

Afghan Special Response Team Round V2

Afghan Special Response Team Round V3

Afghan Special Response Team Round V4

Afghan Special Response Team Rect V1

Afghan Special Response Team Rect V2

Afghan Forces Development Unit Rot 18 V1

Afghan Forces Development Unit V1

Afghan Forces Development Unit V2

Afghan TF666 V1

Afghan TF666 V2

Afghan TF666 V3

Afghan TF666 V4

ISAF — SOF V1

ISAF — SOF V2

ISAF — SOF V3

Afghanistan

ISAF — SOF V4

ISAF — SOF V5

ISAF — SOF V6

ISAF — SOF V7

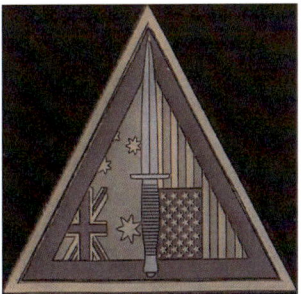
ISAF — SOF AUSTUS (PVC)

ISAF — SOAG

Special Operations Task Force East — Afghan V1

Special Operations Task Force East — Afghan V2

Special Operations Task Force East — Afghan V3

Special Operations Task Force East — Afghan V4

Special Operations Task Force East — Afghan V5

SF Embeds V1

SF Embeds V2

SF Embeds V3

SF Embeds V4

Afghanistan

SF Embeds V5

SF — ANA Commandos V1

SF — ANA Commandos V2

SF — ANA Commandos V3

SF — ANA Commando School

Novelty Special Forces Afghan

AFGHANISTAN

Headquarters

HQ JTF 633-A ISAF Oval V1

HQ JTF 633-A ISAF Oval V2

HQ JTF 633-A ISAF V1

HQ JTF 633-A ISAF V2

HQ JTF 633-A ISAF V2 ANF

HQ JTF 633-A ISAF V3

HQ JTF 633-A ISAF V4

HQ JTF 633-A ISAF V5

HQ JTF 633-A Black Roo on MC V1

HQ JTF 633-A Black Roo on MC V2

HQ JTF 633-A Black Roo on MC V3

HQ JTF 633-A Black Roo on MC V4

HQ JTF 633-A Black Roo on MC V5

JTF 636 Resolute Support MC V1

JTF 636 Resolute Support MC V2

JTF 636 Resolute Support MC V3

JTF 636 Resolute Support MC V4

JTF 636 Resolute Support MC V5

JTF 636 Resolute Support MC V6

JTF 636 Resolute Support V1

JTF 636 Resolute Support V2

Afghanistan

JTF 636 Resolute Support V3

JTF 636 Resolute Support V4 Round

JTF 636 Resolute Support V4

Task Group Afghanistan V1

Task Group Afghanistan V2

Task Group Afghanistan V3

Op Resolute Task Group Afghanistan MC V1

Op Resolute Task Group Afghanistan MC V2

Op Resolute Task Group Afghanistan MC V3

Op Resolute Task Group Afghanistan MC V4

Op Resolute Task Group Afghanistan 7 RAR

Op Resolute Task Group Afghanistan 7 RAR K-042

Op Resolute Task Group Afghanistan MC V5

JTF 636 MA — Resolute Support

JTF 636 MA — Resolute Support

Afghanistan

HQ Resolute Support — Military Police V1

HQ Resolute Support — Military Police V2

JTF 636 Resolute Support Round

ISAF NATO Shield

JTF 636 Resolute Support Shield V1

JTF 636 Resolute Support Shield V2

JTF 636 Resolute Support Shield V3

KABUL ANF/AFGHAN Flags

Novelty Kabul Taxi

JTF 636 Resolute Support Dinasaur

HQ JTF 633-A Novelty V1

HQ JTF 633-A Novelty V2

HQ JTF 633-A Novelty V3

HQ JTF 633-A Novelty V4

Novelty Kabul Gym

Afghanistan

Tactical Driving Team

SIGS Kabul Node V1

SIGS Kabul Node V2

SIGS — ECS-3 Kabul

Kabul Logistics Unit

Afghan National Army Officer Academy V1

Afghan National Army Officer Academy V2

Afghan National Army Officer Academy V3

Afghan National Army Officer Academy V4

Afghan National Army Officer Academy V5

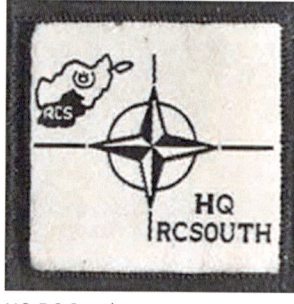
HQ RC South V1

HQ RC South V2

Kabul Garrison Command Advisor Team

Strategic Partnering

Counterinsurgency Training Centre V1

Afghanistan

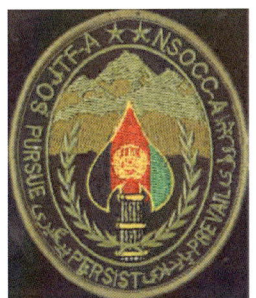

EMBEDS NATO Senior Civilian Representative

Australian Embassy Kabul CPT V1

Australian Embassy Kabul CPT V2

Australian Embassy Kabul CPT V3

Australian Embassy Kabul CPT ANF

AFGHANISTAN

Combined Team Uruzgan

Combined Team Uruzgan Coloured V1

Combined Team Uruzgan Coloured V2

Combined Team Uruzgan Subdued V1

Combined Team Uruzgan Subdued V2

Combined Team Uruzgan Green

CTU ASM

Combined Team Uruzgan — CJ 2 Cell

CTU Public Affairs Cell — TK

CTU Team Ninja Fish

AFGHANISTAN

Force Support Unit

FORCE SUPPORT UNIT RAEME

AFGHANISTAN

Rotary Wing Group

JTF 633.7 Op Brahmans Oval V1

JTF 633.7 Op Brahmans Oval V2

JTF 633.7 Op Brahmans Oval V3

JTF 633.7 Op Brahmans Oval V4

JTF 633.7 Op Brahmans Oval V5

JTF 633.7 Op Brahmans Oval V6

JTF 633.7 Op Brahmans Oval V7

JTF 633.7 Op Brahmans Oval V8

JTF 633.7 Op Brahmans Round

JTF 633.7 Op Brahmans Oval Coloured V1

JTF 633.7 Op Brahmans Oval Coloured V2

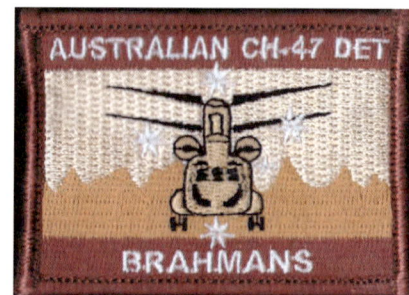
JTF 633.7 Op Brahmans V1

JTF 633.7 Op Brahmans V2

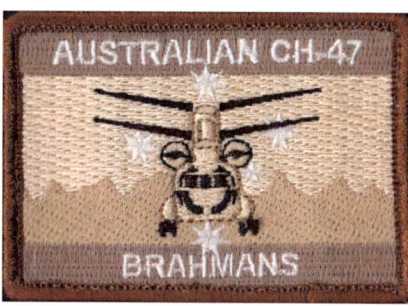
JTF 633.7 Op Brahmans V3

JTF 633.7 Op Brahmans V4

JTF 633.7 Op Brahmans V5

JTF 633.7 Op Brahmans V6

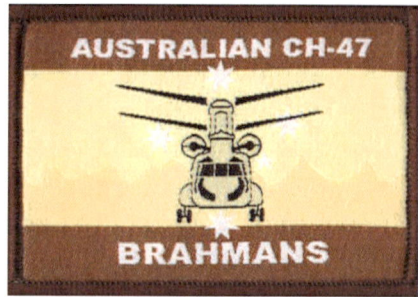
JTF 633.7 Op Brahmans V7

Afghanistan

JTF 633.7 Op Brahmans V8

JTF 633.7 Op Brahmans V9

JTF 633.7 Op Brahmans Night Watchman

JTF 633.7 Op Brahmans 5 Aviation Regt V1

JTF 633.7 Op Brahmans 5 Aviation Regt V2

JTF 633.7 C Sqn 5 Aviation Regt Coloured V1

JTF 633.7 C Sqn 5 Aviation Regt Coloured V2

JTF 633.7 C Sqn 5 Aviation Regt Subdued V1

JTF 633.7 C Sqn 5 Aviation Regt Subdued V2

JTF 633.7 Op Brahmans C Sqn OPS V1

JTF 633.7 Op Brahmans C Sqn OPS V2

JTF 633.7 Op Brahmans 23 Combat Techo

JTF 633.7 Op Brahmans RAEME

Aviation Chinook Aircrew

Aviation Chinook CH-47

Afghanistan

Aviation Chinook Loadmaster

JTF 633.7 Op Brahmans Loadmaster Mafia

JTF 633.7 Airframe 1 A15-202

JTF 633.7 Airframe 2 A15-104

JTF 633.7 Airframe 3 A15-106

JTF 633.7 Airframe 4 A15-201

JTF 633.7 Airframe 5 A15-103 Crashed

JTF 633.7 Airframe 6 A15-102 Crashed

AFGHANISTAN

Artillery

Artillery Det

Artillery Training Advisory Team Round

Artillery Training Advisory Team V1

Afghanistan

Artillery Training Advisory Team V2

Artillery Training Advisory Team V1 Coloured

Artillery Training Advisory Team V1 Subdued

Artillery Training Advisory Team V2 Coloured

Artillery Training Advisory Team V2 Subdued

AFGHANISTAN

Coalition Advisory Team

205th Hero Corps CAT V1

205th Hero Corps CAT V2

205th Hero Corps CAT V3

205th Hero Corps CAT V4

205th Hero Corps CAT V5

205th Hero Corps CAT V6

205th Hero Corps CAT Subdued

205th Hero Corps CAT COLF 5

205th Hero Corps B Coy 1 RAR KAF

205th Hero Corps 1 RAR 5 Pl KAF

205th Hero Corps 1 RAR 6 Pl KAF

205th Hero Corps 1 RAR C Coy 8 Pl KAF V1

205th Hero Corps 1 RAR C Coy 8 Pl KAF V2

205th Hero Corps 1 RAR C Coy 8 Pl KAF V3

205th Hero Corps 8 Pl 3/4 Cav Regt KAF

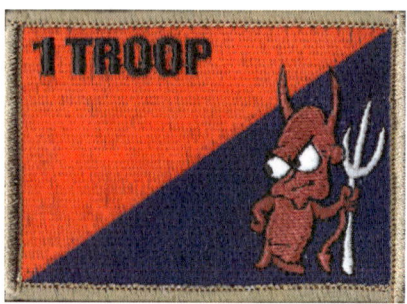
205th Hero Corps 1 Tp 3/4 Cav Regt KAF V1

205th Hero Corps 1 Tp 3/4 Cav Regt KAF V2

Guardian Angels

Afghanistan

AFGHANISTAN

Medical

Medical AUSMTF 3

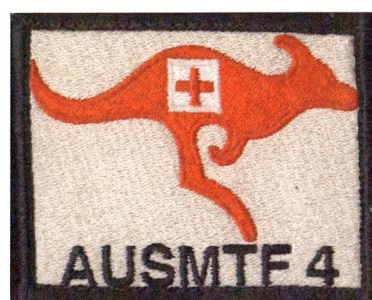
Medical AUSMTF 4

Medical ASHG — KAF Role 3 V1

Medical ASHG — KAF Role 3 V2 Sample V1

MedicalL ASHG — KAF Role 3 V2 Sample V2

Medical ASHG — KAF Role 3 V2

Medical ASHG — KAF Role 3 V3

Medical HKIA R2E Medical Team (MC)

Medical HKIA R2E Medical Team (Tan)

Medical AME Team 2016

Medical AME Team

Medical ASHG — KAF Role 3 Shield

Afghanistan

Medical — EVACISTAN

Kandahar Air Base 651 EAES

RAAMC EMBEDS in US Unit V1

RAAMC EMBEDS in US Unit V2

RAAMC EMBEDS in US Unit V3

RAAMC EMBEDS in US Unit V4

RAAMC EMBEDS in US Unit V5

RAAMC EMBEDS in US Unit V6

RAAMC EMBEDS in US Unit V7

AFGHANISTAN

UAV

TG 633.12 Taipan Heron Detachment V1

TG 633.12 Taipan Heron Detachment V2

TU 633.2.7 Aust Heron Det V1

TU 633.2.7 Aust Heron Det V2

TU 633.2.7 AAC Heron UAV V1

TU 633.2.7 AAC Heron UAV V2

TU 633.2.7 AAC Heron UAV V3

Heron Detachment Rot 7

TG633.2.7 Heron Rot 8 'Knobs' 2011

TG633.2.7 Rot 8 2011

TU 633.2.7 AUST HERON DET 11

TG633.2.7 Heron Detachment Rot 12 Devil's Dozen 2013

Heron Detachment Rot 13

Heron Detachment Rot 15

Heron Detachment

UAV Scan Eagle

Afghanistan

AFGHANISTAN
RTF MRTF MTF

Recontruction Task Force 1 Coloured

RTF 1 Subdued

RTF 1 21 Const Regt Coloured

RTF 1 21 Const Regt Subdued

RTF 1 21 Const Regt Resources Tp

RTF 2 V1

RTF 2 V2

RTF 2 1 Tp B Sqn 3/4 Cav Regt

RTF 2 3 RAR Wings

RTF 2 RAEME

RTF 3

RTF 3 Fake

Afghanistan

RTF 3 7 Pl C Coy Unforgiven

RTF 4

Mentoring Recontruction Task Force 1

MRTF 1 CO's Tac Party Coloured

MRTF 1 CO's Tac Party Subdued

MRTF 1 RASIGS

MRTF 2 Sample

MRTF 2 Unoffical

MRTF 2 2 Cav Regt

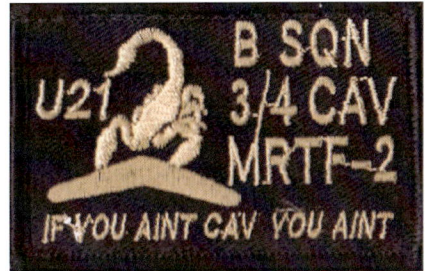
MRTF 2 B Sqn 3/4 Cav Regt

MRTF 2 C-S 21 Search Tp V1

MRTF 2 C-S 21 Search Tp V2

MRTF 2 16 Tp 3 CER

MRTF 2 Engineers 6 ESR

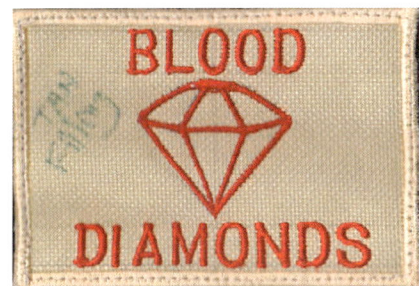
MRTF 2 Engineers V1

Afghanistan

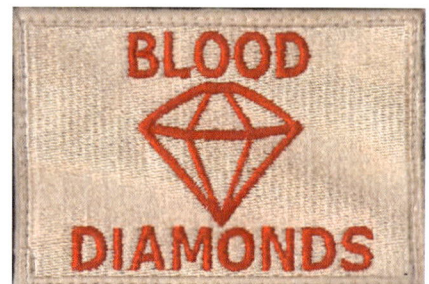
MRTF 2 Engineers V2

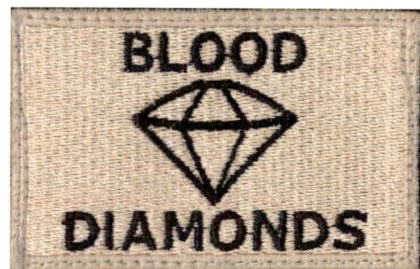
MRTF 2 Engineers V3

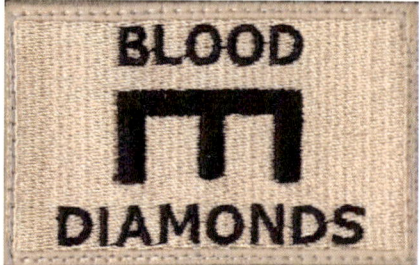
MRTF 2 Engineers V4

MRTF 2 Engineers V5

MRTF 2 Engineers V6

MRTF 2 PDD

MRTF 2 1 RAR Mortar Pl V1

MRTF 2 1 RAR Mortar Pl V2

MRTF 2 Wksps

MRTF_2 Wksps Recovery Mechanic

MRTF 2 OMLT V1

MRTF 2 OMLT V2

MRTF 2 OMLT Name Tag

MRTF 2 OMLT Interpreter

MRTF 2 OMLT Interpreter Flag V1

Afghanistan

MRTF 2 OMLT Interpreter Flag V2

MRTF 2 1 RAR Pineapple

Mentoring Task Force 1 — 6 RAR BN GP

MTF 1 CT-D Coloured (PVC)

MTF 1 CT-D Subdued (PVC)

MTF 1 Ironside 42

MTF 1 OMLT — D 42

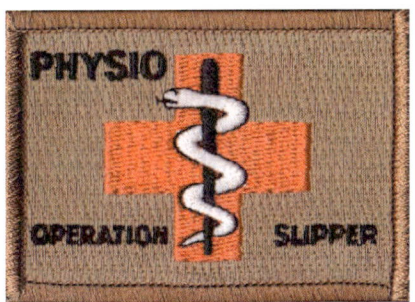
MTF 1 Physio

MTF 2 Team Alpha Coloured

MTF 2 Team Alpha Subdued

MTF 2 CT Bravo

MTF 2 CT-C 5 RAR V1

MTF 2 CT-C 5 RAR V2

MTF 2 CT-C 5 RAR V3

MTF 2 CT-C 5 RAR V4

Afghanistan

MTF 2 Callsign 62

MTF 2 Mortar Pl 5 RAR V1

MTF 2 Mortar Pl 5 RAR V2

MTF 2 Mortar Pl 5 RAR V3

MTF 2 CBT Engineers

MTF 2 CT Engineers 15 Tp

MTF 2 Engineers V1

MTF 2 Engineers V2

MTF 2 Engineers V3

MTF 2 Engineers V4

MTF 2 Engineers V5

MTF 2 Engineers V6

MTF 2 EOD V1

MTF 2 EOD V2

MTF 2 EOD V3

Afghanistan

MTF 2 EOD V4

MTF 2 EOD V5

MTF 2 Combat Medic V1

MTF 2 Combat Medic V2

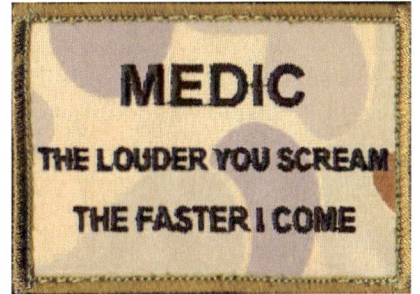
MTF 2 Medic DPDU

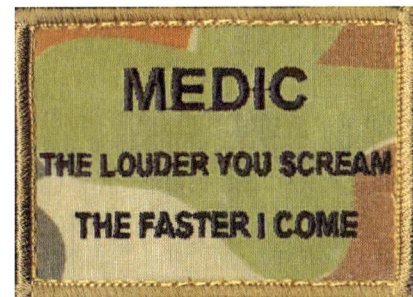
MTF 2 Medic DPCU

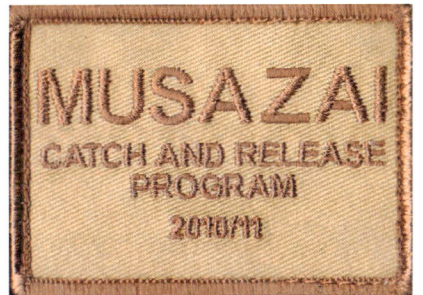
MTF 2 Paltrol Base Musazai

MTF 2 UAV GP V1

MTF 2 UAV GP V2

MTF 2 UAV GP 8

MTF 2 UAV GP 25000 HRS

MTF 2 Bushmaster Taxi Service

MTF 3 2 Tp B Sqn 3/4 Cav Regt

MTF 3 C Sqn 2/14 LHR V1

MTF 3 C Sqn 2/14 LHR V2

Afghanistan

MTF 3 2 RAR 2 Pl

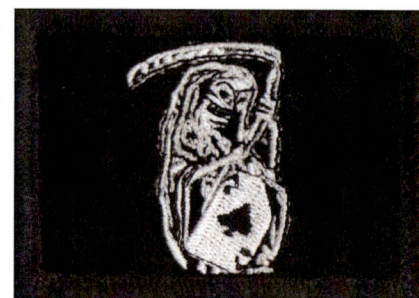
MTF 3 2 RAR Mortar Pl V1

MTF 3 2 RAR Mortar Pl V2

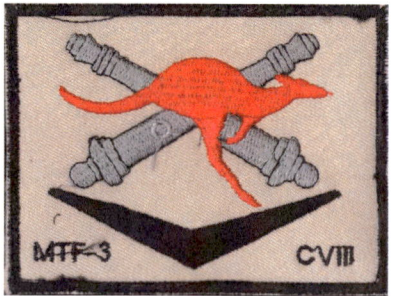
MTF 3 JTAC 108 Fd Bty 4 Regt V1

MTF 3 JTAC 108 Fd Bty 4 Regt V2

MTF 3 EOD Wallaby 2 V1

MTF 3 EOD Wallaby 2 V2

MTF 3 EOD Wallaby 3 V1

MTF 3 EOD Wallaby 3 V2

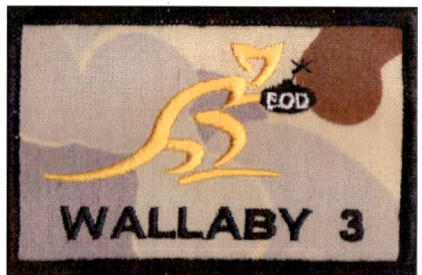
MTF 3 EOD Wallaby 3 V3

MTF 3 72 EW Sqn RAVEN

MTF 3 Weapons Intelligence Team 5

MTF 3 RAEME

MTF 3 Wksps

MTF 3 UAV GP DPDU

Afghanistan

MTF 3 UAV GP DPCU

MTF 3 Dhobi No TBAS Blank

MTF 3 CT-C Dhobi No

MTF 3 CT-C Dhobi ANF

MTF 3 Provincial Recontructional Team OGA

MTF 3 Provincial Reconstruction Team

MTF 3 Interpreters Flag V1

MTF 3 Interpreters Flag V2

MTF 3 Interpreters Flag V3

MTF 3 Interpreters Flag V4

MTF 3 Novelty

MTF 3 Novelty

MTF 4 A Sqn 2/14 LHR

MTF 4 3 RAR Heavy Weapons Pl

MTF 4 3 RAR Signal Pl

Afghanistan

MTF 4 3 RAR Wings V1

MTF 4 3 RAR Wings V2

MTF 4 Mortar Pl Callsign 80 MC

MTF 4 Mortar Pl Callsign 80 DPCU

MTF 4 Mentoring Team 1 V1

MTF 4 Mentoring Team 1 V2

MTF 4 Mentoring Team 1 V3

MTF 4 EDO 20.2

MTF 4 JTAC Stonecutters

MTF JTAC Gunners

MTF 4 Dhobi No

MTF 4 Novelty

MTF 5 2 Cav Task Gp V1

MTF 5 2 Cav Task Gp V2

MTF 5 2 Cav Task Gp B Sqn

Afghanistan

MTF 5 2 Cav Task Gp 1 Tp B Sqn V1

MTF 5 2 Cav Task Gp 1 Tp B Sqn V2

MTF 5 Mentoring Team 4 V1

MTF 5 Mentoring Team 4 V2

MTF 5 EOD

MTF 5 Novelty

RAEME — FIT Team

RAEME IIS Team Project Bushmaster

RAEME Project Bushmaster

Conter Rocket Artillery & Mortar Missile

FSU Camp Maintenance Team — TK

Force Support Unit FST — TK

Non Govt Organisation

Unknown OMLT Team 1

Weapons Intelligence Team Operator V1

Afghanistan

Weapons Intelligence Team Operator V2

US Fire Service TK

TQ Military Police

AFGHANISTAN

Multi National Base Command — TK

MNBC — TK V1

MNBC — TK V2

MNBC — TK V3

MNBC — TK V4

MNBC — TK V5

MNBC 2 Airfield Defence Guard V1

MNBC 2 Airfield Defence Guard V2

MNBC 2 Airfield Defence Guard V3

MNBC Novelty (MC)

MNBC — TK RAAF Ensign

AFGHANISTAN

National Emergency Operation

National Emergency Operation 2021 V1

National Emergency Operation 2021 V2

AFGHANISTAN

Misc

ISAF Round Green V1

ISAF Round Green V2

ISAF Round Green V3

ISAF Round Tan V1

ISAF Round Tan V2

ISAF Round Tan V3

ISAF Round MC

ISAF Rect DPCU V1

ISAF Rect DPCU V2

ISAF Rect DPDU

ISAF Rect MC V1

ISAF Rect MC V2

ISAF Rect MC V3

ISAF Rect MC V4

ISAF Rect MC V5

ISAF Rect Tan V1

ISAF Rect Tan V2

ISAF Green

Afghanistan

ISAF NATO Shield V1

ISAF NATO Shield V2

ISAF NATO Shield V3

ISAF NATO Shield V4

ISAF Shield NTM-A MC

ISAF SOF NATO Shield V1

ISAF SOF NATO Shield V2

ISAF Scroll

Nametag

Dhobi No A042

Dhobi No A056

Dhobi No CAM897

Dhobi No I307

Dhobi No J90N

Dhobi No KB824

Afghanistan

Dhobi No KM379

Dhobi No O34

Dhobi No R601

Novelty Operation Slipper — ARMY

Novelty Operation Slipper — RAN

Novelty Afghan Ribbon Bar Tarin Kowt 2009

Novelty Afghan Ribbon Bar

ANF AFG Tarin Kowt 2011-12

Aust/Afghan Flag

Novelty Taliban Hunting Club

Afghan Ministry of Interior Affairs

ISAF Coalition Dustoff

Novelty ISAF Four Corners

Novelty INFIDEL V1

Novelty INFIDEL V2

Afghanistan

Novelty INFIDEL V3

Novelty Major League Infidel V1

Novelty Major League Infidel V2

Novelty Major League Infidel V3

Novelty Major League Sniper

Novelty Major League Door Kicker V1

Novelty Major League Door Kicker V2

Novelty Major League Cavalry

Novelty Major League Searcher

Novelty Major League Doghandler

Novelty Major League Taliban Hillfighter

Novelty OEF

Novelty Punisher V1

Novelty Punisher V2

Novelty Punisher V3

Afghanistan

Novelty Sons of Anarchy V1

NOVELTY Sons of Anarchy V2

Novelty Achmed

Novelty The Land of the AK 47

Novelty Hard Rock Cafe

Novelty Basic Food Groups

Novelty Guns & Coffee

Novelty Hadji Don't Surf

Novelty Fuck Bin Laden

Novelty Harley Davidson Uruzgan Chapter

Unknown 1 OMLT

Unknown 2

Afghanistan

AUSTRALIAN OPERATIONS

JTF 114 Op Gold Sydney Olympics 2000

JTF 114 Op Gold Sydney Olympics

JTF 114 F Troop 2 Sqn SASR Op Gold Sydney Olympics

AUSTRALIAN OPERATIONS

JTF 629 COVID-19 Assist

JTF 629 COVID-19 Assist (Army)

JTF 629 COVID-19 Assist (Army) Subdued V1

JTF 629 COVID-19 Assist (Army) Subdued V2

JTF 629 COVID-19 Assist (RAN)

JTF 629 COVID-19 Assist (RAAF)

JTU 629.2.2 COVID-19 Assist (RAAF)

JTU 629.2.2 COVID-19 Assist (RAN)

JTF 629 5 Brigade TU Brahman

JTG 629.7 COVID Assist NT

JTF 629.7.4 4 Brigade Shrike

JTF 629 ADF WA 2020

Novelty OP Enduring Clusterfuck

Novelty OP Enduring Fuckery

AUSTRALIAN OPERATIONS

JTF 634 Op Deluge

JTF 634 Op Deluge

AUSTRALIAN OPERATIONS

JTF 635 Op Acolyte Melb C'Wealth Games 2006

JTF 635 Op Acolyte Melb C'Wealth Games Round

JTF 635 Op Acolyte Melb C'Wealth Games

JTF 635 Op Acolyte C Coy 4 RAR

JTF 635 Op Acolyte 4 Brigade Reserve Response Force

AUSTRALIAN OPERATIONS

JTF 637 Op Atlas Gold Coast C'Wealth Games 2018

JTF 637 Op Atlas Gold Coast C'Wealth Games (Army)

JTF 637 Op Atlas Gold Coast C'Wealth Games (RAN)

JTF 637 Op Atlas Gold Coast C'Wealth Games (RAAF)

AUSTRALIAN OPERATIONS

JTF 637 Op Queenslander

JTF 637 Op Queenslander

AUSTRALIAN OPERATIONS

JTF 639 Op Resolute Border Protection

JTF 639 Op Resolute TSE 54

JTF 639 Op Resolute TSE 59

JTF 639 Op Resolute TSE 60

JTF 639 Op Resolute TSE 61

JTF 639 Op Resolute TSE 62

JTF 639 Op Resolute TSE 63

JTF 639 Op Resolute TSE 72 (Army)

JTF 639 Op Resolute TSE 72 (RAN)

JTF 639 Op Resolute TSE 74 (Army)

R10JTF 639 Op Resolute TSE 74 (RAN)

JTF 639 Op Resolute 51 FNQR

JTF 639 Op Resolute Norforce

JTF 639 Op Resolute Pilbara Regt

JTF 639 Op Resolute RFSU

JTF 639 Op Resolute TSE

JTF 639 Op Resolute TSE

JTF 639 Op Resolute TSE (RAN)

JTF 639 Op Resolute

JTF 639 Border Protection Command (Army)

JTF 639 Border Protection Command (RAN)

JTF 639 Border Protection Command (RAAF)

Australian Operations

JTF 639 Op Resolute RFSU Force Element

JTF 639 Op Resolute C/S 84 Starboard Watch Black V1

JTF 639 Op Resolute C/S 84 Starboard Watch Black V2

JTF 639 Op Resolute C/S 84 Starboard Watch Tan

TG 639

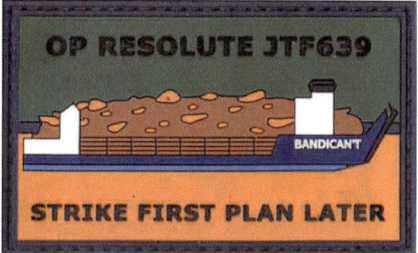
JTF 639 Op Resolute Strike First Plan Later PVC

JTF 639 Op Resolute (Army)

JTF 639 Op Resolute (Army) (Subdued) V1

JTF 639 Op Resolute (Army) (Subdued) V2

JTF 639 Op Resolute (Army) (Subdued) V3

JTF 639 Op Resolute (Army) (Subdued) V4

JTF 639 Op Resolute (RAAF)

JTF 639 Op Resolute (RAAF) (Subdued)

JTF 639 Op Resolute (Tri Service)

JTF 639 Op Resolute (Tri Service) (Subdued)

Australian Operations

JTG 639.2 Op Resolute

JTG 639.2 Op Resolute (Subdued)

JTGF 639.2.4 Op Resolute

JTU 639.2.4 Op Resolute (Subdued)

SF TAG East Rot 7 Round

SF TAG East Rot 7

JTF 639 Op Overarch

AUSTRALIAN OPERATIONS

JTF 640 Op Norwich

JTF 640 Op Norwich

AUSTRALIAN OPERATIONS

JTF 641 Op Outreach

JTF 641 Op Outreach

AUSTRALIAN OPERATIONS

JTF 644 Op Amulet

JTF 644 Op Amulet

Australian Operations

AUSTRALIAN OPERATIONS

JTF 646 Op Invincible Invictus Games 2018

JTF 646 Op Invincible 18 Invictus Games (Army)

JTF 646 Op Invincible 18 Invictus Games (RAN)

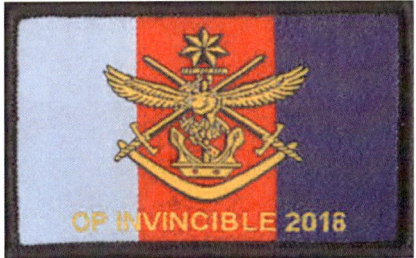
JTF 646 Op Invincible 18 Invictus Games (RAAF)

AUSTRALIAN OPERATIONS

JTF 662 Operation Vic Fire Assist 2009

JTF 662 Operation Vic Fire Assist — FORENSICS

JTF 662 Victorian Bushfire Tragedy Response (Army)

JTF 662 Victorian Bushfire Tragedy Response — New Life New Hope

JTF 662 Victorian Bushfire Survivor Donation Black

JTF 662 Victorian Bushfire Survivor Donation Red

JTF 662 Victorian Bushfire Survivor Donation Yellow

AUSTRALIAN OPERATIONS

JTF 664

JTF 664.1

AUSTRALIAN OPERATIONS

JTF 664 Op Cyclone Assist 2017 (TC Debbie)

JTF 664 OP Cyclone Assist   JTF 664 Queensland Reconstruction Team

AUSTRALIAN OPERATIONS

JTF 665 Op Parapet

JTF 665 Op Parapet

AUSTRALIAN OPERATIONS

JTF 665 Op Testament

JTF 665 OP TESTAMENT

AUSTRALIAN OPERATIONS

Op Bushfire Assist 2019-20

JTF 646.2 Op Bushfire Assist 19/20 V1

JTF 646.2 Op Bushfire Assist 19/20 V2

JTF 646.2 Engineers Callsign 12

JTF 646.2 Op Bushfire Assist 11 Brigade Element

JTF 646.2 Op Bushfire Assist FSG 17 Brigade Element

JTF 646.2 Op Bushfire Assist 6 ESR Element

JTF 646.2 6 Op Bushfire Assist 6 ESR Slow Down

JTF 646.2 DCC and Support Crew

JTF 1110 Op Bushfire Assist 2020 It's How We Roll

JTF 1110 Op Bushfire Assist 2020 Red Shit Happens

35 Sqn (RAAF) Element

INDO-PACIFIC REGION

Bougainville

Bougainville V1

Bougainville V2

Bougainville AFP Brassard

INDO-PACIFIC REGION

CJTF 629 Op Sumatra Assist 2004

CJTF 629 Op Sumatra Assist 2004 V1.JPG

CJTF 629 Op Sumatra Assist 2004 V2.JPG

INDO-PACIFIC REGION

Indo Pacific Endeavour

IPE 18 JTG Indo Pacific Endeavour

IPE 21 JTG Indo Pacific Endeavour (Army)

IPE 21 JTG Indo Pacific Endeavour (RAN)

IPE 21 JTG Indo Pacific Endeavour (RAAF)

INDO-PACIFIC REGION

JTF 629 Op Samoa Assist 2009

JTF 629 Op Samoa Assist

INDO-PACIFIC REGION

JTF 630 Op Pacific Assist

JTF 630 Op Pacific Assist

INDO-PACIFIC REGION

JTF 637 Pacific Islands Support

Pacific Partnership (Army)

Pacific Partnership (RAN)

JTF 637 Pacific Islands Support (Army)

JTF 637 Pacific Islands Support (RAN)

JTF 637 Pacific Islands Support (RAAF)

Pacific Support Team (Army)

Pacific Support Team (RAN)

Pacific Support Team (RAAF)

INDO-PACIFIC REGION

JTF 658 MH370 International Search Operation Southern Indian Ocean

JTF 658 MH370 International Search Operation Southern Indian Ocean Round V1

JTF 658 MH370 International Search Operation Southern Indian Ocean Round V2

JTF 658 MH370 International Search Operation Southern Indian Ocean

JTF 658 MH370 Search & Rescue Team Large

JTF 658 MH370 Search & Rescue Team Small

JTF 658 MH370 Search & Rescue Team — Perth

JTF 658 MH370 Underwater Search

JTF 658 KR795 SEARCH FOR MH370

LIMA 2017 MH370 Search & Rescue Exercise

Indo-Pacific Region

INDO-PACIFIC REGION

Malaysia

Australian Embassy Kuala Lumur

Australian Embassy Kuala Lumur (Subdued)

Australian Embassy Kuala Lumur AMCU

Australian Embassy Kuala Lumur AMCU (Subdued)

Archaios Battlefield Search & Secure

Archaios Battlefield We Bring Them Home

2/30 Trg Gp Rifle Company Butterworth

2/30 Trg Gp Rifle Company Butterworth COVID-19

2/30 Trg Gp Rifle Company Butterworth

Rifle Company Butterworth

Rifle Company Butterworth Rot 84

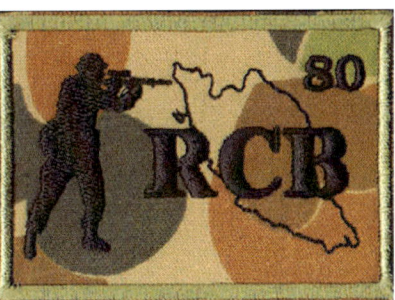
Rifle Company Butterworth Rot 80

Rifle Company Butterworth Rot 87

Rifle Company Butterworth Rot 99 — 3 RAR V1

Rifle Company Butterworth Rot 99 — 3 RAR V2

Rifle Company Butterworth Rot 111 — B Sqn 2/14 LHR

Rifle Company Butterworth Rot 128

Rifle Company Butterworth Rot 134 — Novelty Autism Subdued

Rifle Company Butterworth Rot 130 — 1 Aviation Regt

Rifle Company Butterworth Rot 132

Rifle Company Butterworth Rot 134 — Novelty Autism Subdued

Rifle Company Butterworth Rot 134 — 6 RAR

Rifle Company Butterworth Rot 134 — A Coy 6 RAR

Rifle Company Butterworth Rot 134

Rifle Company Butterworth Rot 134 — Novelty Autism Coloured

Rifle Company Butterworth Rot 136

Rifle Company Butterworth Rot 136 — Novelty Autism

Indo-Pacific Region

Novelty Boxing Kangaroo

Novelty Tiger Beer

PNG Flag AMCU

Panthers R.F.C.

ADFTCC

Aboriginal Flag

2 Sqdn (RAAF)

13 Sqn (RAAF)

19 Sqn Butterworth (RAAF)

92WG Det A (RAAF) GC

92WG Det A (RAAF) AC

92WG Det A (RAAF) V1

92WG Det A (RAAF) V2

324 Combat Support Sqn

Exercise Stardex 1999 Nimrod Ground Crew

Indo-Pacific Region

Exercise Bersama Shield 2005 — Monkey Love Tour

Exercise Bersama Lima 2005 — Lager Lovers Tour

Exercise Bersama Lima 16 2 Sqn

Unknown Crew 3

RAN Unknown Malaysia Tour

INDO-PACIFIC REGION

Papau New Guinea

CJTF 630 PNG

JTG 658 OP Hannah PNG Elections

PNG 3 CER EX Puk Puk

Op APEC 18 Assist (Army)

Op APEC 18 Assist (RAN)

Op APEC 18 Assist (RAAF)

PNG Defence Cooperation Program (Army) V1

PNG Defence Cooperation Program (RAN) V1

PNG Defence Cooperation Program (RAAF) V1

PNG Defence Cooperation Program (Army) V2

PNG Defence Cooperation Program (RAN) V2

PNG Defence Cooperation Program (RAAF) V2

PNG Defence Cooperation Program (Army) V3

PNG Defence Cooperation Program (RAN) V2

PNG Defence Cooperation Program (RAAF) V3

PNG Mobile Training Team Charlie

PNG Mobile Training Team Charlie V2

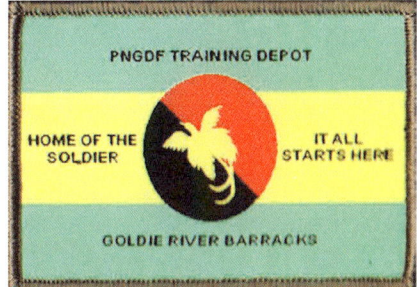
PNGDF Training Depot

PNGDF Training Depot (Subdued)

Indo-Pacific Region

INDO-PACIFIC REGION
Phillippines

TU 629.3.1 Op Augury Round

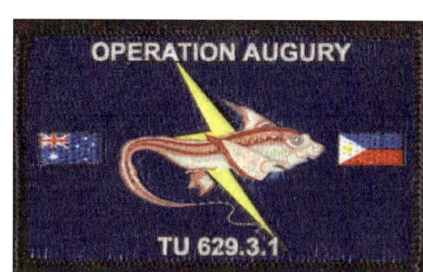
TU 629.3.1 Op Augury Rect

JTG 629 Op Augury (Army) V1

JTG 629 Op Augury (RAN) V1

JTG 629 Op Augury (RAAF) V1

JTG 629 Op Augury (Army) V2

JTG 629 Op Augury (RAN) V2

JTG 629 Op Augury (RAAF) V2

JTG 629 Op Augury Land Mobile Training Team

JTG 629 Op Augury AWM Visit

P11@JTG 629 Op Augury 2 CER V1.JPG

P12@JTG 629 Op Augury 2 CER V2.JPG

P13@JTG 629 Op Augury 2 CER Breach Clear.JPG

P14@JTG 629 Op Augury 2 CER Urban Search & Breach tab.JPG

P15@JTG 629 Op Augury 2 CER Big Daddy Breach Nametag.JPG

P16@JTG 629 Op Augury 2 CER Sapper tab.JPG

P17@JTG 629 Op Augury 2 CER Combat Life Safer tab.JPG

P18@JTG 629 Op Augury 2 CER Major League Searcher.JPG

P19@JTG 629 Op Augury 2 CER IR ANF V1.JPG

P20@JTG 629 Op Augury 2 CER IR ANF V2.JPG

P21@JTG 629 Op Augury 2 CER Phillipines Flag.JPG

INDO-PACIFIC REGION

Solomons

CTF 635 Op Anode Solomon Islands V1

CTF 635 Op Anode Solomon Islands V2

CTF 635 Op Anode Solomon Islands V3

JTF 663 Op Render Safe V2

JTF 663 Op Render Safe V3

CTG 634 Solomon Islands Elections 2019 (Army)

CTG 634 Solomon Islands Elections 2019 (RAN)

CTG 634 Solomon Islands Elections 2019 (RAAF)

JTG 637.3 Solomons' International Assisitance Force V1

JTG 637.3 Solomons' International Assisitance Force V2

JTG 637.3 Aust/Solomons Flags

INDO-PACIFIC REGION

Timor Leste

INTERFET Title

INTERFET Combined Airlift Wing

INTERFET International Combined Airlift Wing

Indo-Pacific Region

United Nations V1

United Nations V2

United Nations V3

United Nations Round 4 RAR (Subdued)

2 RAR Assault Pioneers Black

2 RAR Assault Pioneers DPCU

2 RAR Assault Pioneers RCCADTE Black

2 RAR Assault Pioneers RCCADTE DPCU

1 CER

OP Spire 2 2004-05

AUSBATT 7 5/7 RAR BG

AUSBATT 9 6 RAR BG

Timor Leste Op Astute

JTF 631 Timor Leste V1

JTF 631 Timor Leste V2

Indo-Pacific Region

JTF 631 Timor Leste V3

JTF 631 Timor Leste V4

JTF 631 Op Astute

JTF 631 Timor Leste NZ

2 Pl A Coy 8/9 RAR

JTF 631 Timor Leste TLBG Enigeer Gp 2 CER Tan

JTF 631 Timor Leste TLBG Engineer Gp 2 CER DPCU

JTF 631 Timor Leste 08

ANZAC Aviation East Timor

JTF 631 Timor Leste TLAG 15 Blackhawk Crew V1

JTF 631 Timor Leste TLAG 15 Blackhawk Crew V2

JTF 631 Timor Leste Aviation Gp 12

CIMIC Green Brassard

CIMIC V1

CIMIC V2

Indo-Pacific Region

CIMIC V3

CIMIC V4 Indonesian Text

CIMIC V5

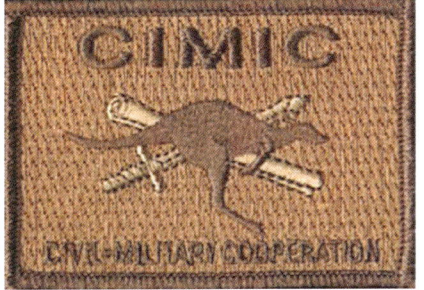
CIMIC V5 Subdued

T038@CIMIC V6.JPG

JTF 631 Timor Leste ATST Green

JTF 631 Timor Leste ATST Large

JTF 631 Timor Leste ATST Small

JTF 631 Timor Leste ATST V1

JTF 631 Timor Leste ATST V2

JTF 631 Timor Leste ATST V3

JTF 631 Timor Leste DCP-EM Brassard

JTF 631 Timor Leste DCP-TL Brassard

JTF 631 Timor Leste ATST-EM Brassard

JTF 631 Timor Leste ATST SHIELD

Indo-Pacific Region

JTF 631 Timor Leste DCP-EM

JTF 631 Timor Leste FALINTIL FDTL

JTF 631 Timor Leste CENTRO DE INSTRUCAO FDTL

JTF 631 Timor Leste F-FDTL

Timor Leste Training UCP Large

Timor Leste Training UCP Small

JTF 631 Defence Cooperation Program East Timor

JTF 631 Defence Cooperation Program Timor Leste V1

JTF 631 Defence Cooperation Program Timor Leste V2

JTF 631 DEFENCE COOPERATION PROGRAM TIMOR LESTE V3

JTF 631 Timor Leste Blackhawk UH-60

JTF 631 Timor Leste Blackhawk Aircrew V1

JTF 631 Timor Leste Blackhawk Aircrew V2

JTF 631 Timor Leste Blackhawk Loadmaster

East Timor Unknown

Indo-Pacific Region

UNMISET Novelty

International Stabilization Force Timor Leste Novelty

INDO-PACIFIC REGION

Tonga

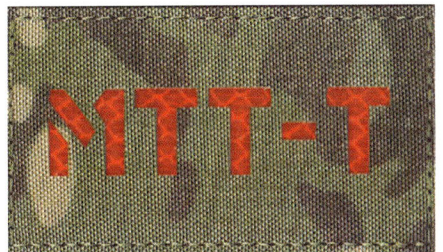
Mobile Training Team — Tonga

INDO-PACIFIC REGION

TU 659 Op Argo

TG 659 UNSCR 2397

TU 659.1 Op Argo AC V1

TU 659.1 Op Argo AC V2

TU 659.1 Op Argo AC V3

TU 659.1 Op Argo GC

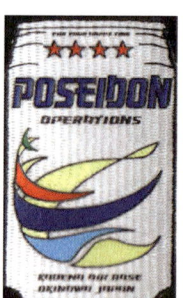
TU 659.1 OP ARGO Posedon Operations

INDO-PACIFIC REGION

Misc

Australian Staff US Central Command

Australian Defence Force South East Singapore

Op Gateway 40 years

HMAS ANZAC Ops Crew

Operation Southern Discovery

IRAQ

Operation Bastille

TF 64 OP BASTILLE CSSG V2

TF 64 Operation Bastille CSSG V1

IRAQ

Operation Falconer

Operation Falconer V1

Operation Falconer V2

IRAQ

Operation Catalyst — Headquarters

HQ JTF 633 V1

HQ JTF 633 V2

Multinational Division (South East)

Bagdad International Airport Tower Controller

Sector Control Point Bagdad

Joint Personnel Recovery Center V1

Joint Personnel Recovery Center V2

Night Stalkers V1

Night Stalkers V2

Night Stalkers V3

Night Stalkers V4

EMBEDS IGFC 10 Div V1

EMBEDS IGFC V1　　　　　EMBEDS IGFC V2　　　　　EMBEDS V1

EMBEDS V2　　　　　　　EMBEDS V3　　　　　　　EMBEDS V4

EMBEDS V5　　　　　　　EMBEDS V6　　　　　　　EMBEDS V7

EMBEDS V8　　　　　　　EMBEDS V9　　　　　　　EMBEDS V10

EMBEDS V11　　　　　　EMBEDS V12　　　　　　EMBEDS V13

EMBEDS V14

EMBEDS V15

EMBEDS V16

EMBEDS V17

EMBEDS V18

EMBEDS V19

EMBEDS V20

EMBEDS V21

EMBEDS V22

EMBEDS V23

EMBEDS V24

EMBEDS V25

EMBEDS V26

IRAQ

IRAQ

Operation Catalyst — Engineers

CEXC Forensics V1

CEXC Forensics V2

CEXC V1

CEXC V2

CEXC V3

CEXC V4

CEXC V5

EOD V1

EOD V2

EOD V3

Survey Group 2003

CMPC Baghdad 2003

CMPC Baghdad 2006-7

EOD Task Force Troy V1

EOD Task Force Troy V2

EOD Task Force Troy V3

EOD Task Force Troy V4

EOD Task Force Troy V5

Mine Information Coordination Cell

Operation Iraqi Freedom EOD V1

Operation Iraqi Freedom EOD V2

J2 Cell Task Force Troy

Task Force Troy Novelty

EOD Novelty V1

EOD Novelty V2

EOD Novelty V3

IRAQ

IRAQ

Operation Catalyst — AATTI

AATTI-SF Counterinsurgency Det

JTF 633 Army Training Team Subdued

JTF 633 Army Training Team Coloured

JTF 633 Army Training Team Coloured Arabic

AATTI Lamassu

Australian Army Training Team Iraq Coloured Round V1

Australian Army Training Team Iraq Coloured Round V2

AATTI Coloured Round Fake

Australian Army Training Team Iraq Coloured Subdued Round V1

Australian Army Training Team Iraq Coloured Subdued Round V2

AATTI Subdued Round Fake

AATTI-5 Coloured

AATTI-5 Coloured Fake

AATTI-5 Subdued

AATTI-5 Subdued Fake

AATTI-6 Coloured

AATTI-7 Coloured DPDU

AATTI-7 Subdued

AATTI-8

AATTI-9 V1

AATTI-9 V2

AATTI-9 V3

AATTI-9 V4

AATTI-9 V5

AATTI Name Tags

IRAQ

IRAQ

Operation Catalyst — SECDET

SECDET 2 Cav Regt

SECDET 7 V1

SECDET 7 V2

SECDET 7 2Tp A Sqn 2/14 LHR

SECDET 8 V1

SECDET 8 V2

SECDET 8 V3

SECDET 9 V1

SECDET 9 V2

SECDET 9 V3

SECDET 10 V1

SECDET 10 V2

SECDET 10 V3

SECDET 10 3 RAR V1

SECDET 10 3 RAR V2

SECDET 10 3 RAR V3

SECDET 10 3 RAR V4

SECDET 10 V32

SECDET 10 V32A

SECDET 10 V32B

SECDET 10 V32C

SECDET 10 V32D

SECDET 10 V32E

SECDET 10 V32F

SECDET 11

SECDET 12 V1

SECDET 12 V2

IRAQ

SECDET 12 V3

SECDET 12 V21 1 Tp B Sqn 2/14 LHR

SECDET 13

SECDET Generic V1

SECDET Generic V2

SECDET V22

IRAQ

Operation Catalyst — AMTG OBG FET

Almuthanna Task Group ASIF

AMTG 1 V1

AMTG 1 V2

UAV Gp Lamussa V1

UAV Gp Lamussa V2

UAV Gp Lamussa V3

UAV Gp Lamussa V4

AMTG 2 V1 20 Armd Inf Bde

AMTG 2 20 Armd Inf Bde ANF

AMTG 2 V2 7 Armd Bde

AMTG 3 Coloured

AMTG 3 Subdued

AMTG 3 Rat on Black ANF

AMTG 3 Rat on Tan ANF V1

AMTG 3 Rat on Tan ANF V2

Japanese Iraq Reconstruction & Support Group (JIRSG) Flag V1

Japanese Iraq Reconstruction & Support Group (JIRSG) Flag V2

OBG(W)-1 V1

OBG(W)-1 V2

OBG(W)-1 V3

OBG(W)-1 V4

OBG(W)-1 V5

OBG(W)-1 V6

OBG(W)-1 V7

OBG(W)-1 Fake V1

OBG(W)-1 Fake V2

OBG(W)-1 BHQ Command Tp

OBG(W)-1 A Sqn 2/14 LHR

OBG(W)-1 CT — Alpha 3 Tp A Sqn 2/14 LHR

OBG(W)-1 6 RAR Bushmaster Crews V1

OBG(W)-1 6 RAR Bushmaster Crews V2

OBG(W)-1 UAV 20 STA DPDU

OBG(W)-2 V1

OBG(W)-2 V2

OBG(W)-2 Fake

OBG(W)-2 141st Airborne Mech Tactical CIMIC Det

IRAQ

OBG(W)-2 Novelty Momento

OBG(W)-3 1 Mech Bde V1

OBG(W)-3 1 Mech Bde V2

OBG(W)-3 V1

OBG(W)-3 V2

OBG(W)-3 V3

OBG(W)-3 Fake

OBG(W)-3 UAV Gp

OBG(W)-3 CT Courage A Sqn 2 Cav Regt

OBG(W)-3 2 Tp B Sqn 3/4 Cav Regt

OBG(W)-3 5 RAR Club Recon DPDU V1

OBG(W)-3 5 RAR Club Recon V2

OBG(W)-4 4 Mech Bde V1

OBG(W)-4 4 Mech Bde V2

OBG(W)-4 4 Mech Bde V3

IRAQ

OBG(W)-4 4 Mech Bde V4

OBG(W)-4 4 Mech Bde V6

OBG(W)-4 4 Mech Bde Fake

OBG(W)-4 4 Mech Bde Tan ANF

OBG(W)-4 V1

OBG(W)-4 V2

OBG(W)-4 V3

OBG(W)-4 V4

OBG(W)-4 V5

OBG(W)-4 V6

OBG(W)-4 Fake V1

OBG(W)-4 Fake V2

OBG(W)-4 Battle Group Chauvel V1

OBG(W)-4 Battle Group Chauvel V2

OBG(W)-4 BHQ S0 Cell

OBG(W)-4 BHQ S1 Cell

OBG(W)-4 BHQ SIG Vault

OBG(W)-4 BHQ Command Tp

OBG(W)-4 BHQ MOVEMENTS CELL

OBG(W)-4 BHQ Legal V1

OBG(W)-4 BHQ Legal V2

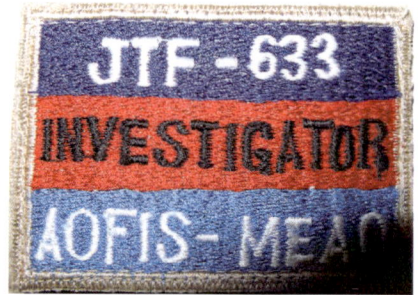
O073@OBG(W)-4 BHQ ADFIS Sample.JPG

OBG(W)-4 BHQ ADFIS V1

OBG(W)-4 BHQ ADFIS V2

OBG(W)-4 BHQ ADFIS V3

OBG(W)-4 BHQ PHYSCOPS

OBG(W)-4 BHQ PHYSCOPS 1 Million Pamphet Drop

OBG(W)-4 BHQ ISF TRG GP V1

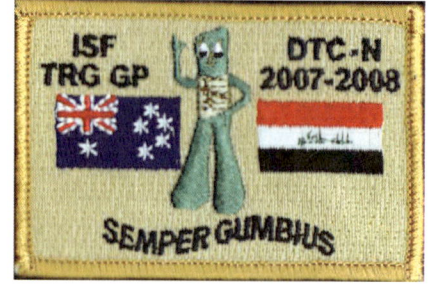
OBG(W)-4 BHQ ISF TRG GP V2

OBG(W)-4 Civilian Police Assistance Training Team Iraq V1

IRAQ

OBG(W)-4 Civilian Police Assistance Training Team Iraq V2

O083@OBG(W)-4 Civilian Police Assistance Training Team Iraq V3.JPG

OBG(W)-4 CIMIC White English

OBG(W)-4 CIMIC White Arabic V1

OBG(W)-4 CIMIC White Arabic V2

OBG(W)-4 CIMIC English DPDU

OBG(W)-4 CIMIC Arabic DPDU

OBG(W)-4 CIMIC Arabic CAM

OBG(W)-4 CT — Heeler V1

OBG(W)-4 CT — Heeler V2

OBG(W)-4 CT — Heeler V3

OBG(W)-4 CT — Heeler V11 1 Tp A Sqn 2/14 LHR V1

OBG(W)-4 CT — Heeler V11 1 Tp A Sqn 2/14 LHR V2

OBG(W)-4 CT — Heeler V11 1 Tp A Sqn 2/14 LHR V3

OBG(W)-4 CT — Heeler A Coy 6 RAR Transformers Bob

OBG(W)-4 CT — Heeler A Coy 6 RAR Transformers Chunks

OBG(W)-4 CT — Heeler A Coy 6 RAR Transformers Gauci

OBG(W)-4 CT — Heeler A Coy 6 RAR Transformers Havo

OBG(W)-4 CT — Heeler A Coy 6 RAR Transformers Hayesy

OBG(W)-4 CT — Heeler A Coy 6 RAR Transformers Joey

OBG(W)-4 CT — Heeler A Coy 6 RAR Transformers Mordog

OBG(W)-4 CT — Heeler A Coy 6 RAR Transformers Mully

OBG(W)-4 CT — Heeler A Coy 6 RAR Transformers Ozzy

OBG(W)-4 CT — Heeler A Coy 6 RAR Transformers Rabbit

OBG(W)-4 CT — Waler V1

OBG(W)-4 CT — Waler V2

OBG(W)-4 CT — Waler Fake

OBG(W)-4 CT — Waler Purple Cobras Coloured

OBG(W)-4 CT — Waler Purple Cobras Subdued

OBG(W)-4 CT — Waler 2 Tp A Sqn 2/14 LHR

IRAQ

OBG(W)-4 CT — Tenacious

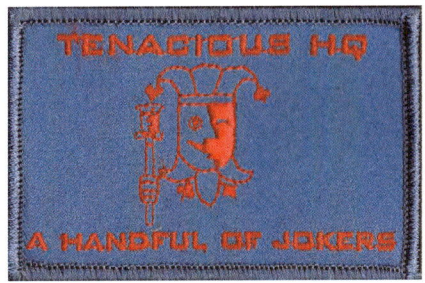
OBG(W)-4 CT — Tenacious HQ

OBG(W)-4 CT — Tenacious RAAOC

OBG(W)-4 CT — Tenacious BGHQ Q Store

OBG(W)-4 CT — Tenacious BGHQ Q Store Not Approved

OBG(W)-4 CT — Tenacious Tpt Tp

OBG(W)-4 CT — Tenacious Arte et Marte Sample

OBG(W)-4 CT — Tenacious Arte et Marte V1

OBG(W)-4 CT — Tenacious Arte et Marte V2

OBG(W)-4 CT — Tenacious RAEME

OBG(W)-4 CT — Tenacious TST

OBG(W)-4 CT — Tenacious TST Recovery Sect

OBG(W)-4 CT — Tenacious Carpenter V1

OBG(W)-4 CT — Tenacious Carpenter V2

O126@OBG(W)-4 CT — Tenacious PTI V1.JPG

IRAQ

OBG(W)-4 CT — Tenacious PTI V2

OBG(W)-4 CT — Tenacious PTI V3

OBG(W)-4 FHT

OBG(W)-4 S Hughes Cell

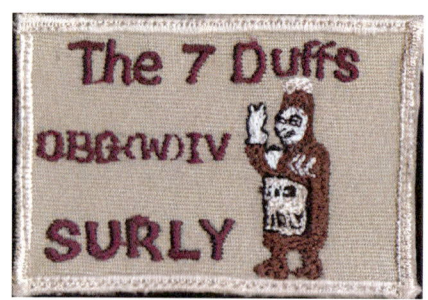
OBG(W)-4 The 7 Duffs Surly

OBG(W)-4 State of Origin Blues Supporter

OBG(W)-4 Aboriginal Flag V1

OBG(W)-4 Aboriginal Flag V2

OBG(W)-4 Aboriginal Flag Error

OBG(W)-4 CGS Potection Party

Tallil Air Base V1

Tallil Air Base V2

CIMIC Brassard DPDU

FET Prototype V1

FET Prototype V2

IRAQ

FET Sample V1

FET Sample V2

FET Coloured

FET Subdued

FET Emergency Response

FET HAT

FET Movements Cell

FET K Rudd Get Out of Iraq

ASLAV Enhancement Team

Unknown 1

IRAQ

IRAQ

Operation Catalyst — Novelty

Novelty OIF V1

Novelty OIF V2

Novelty OIF V3

Novelty OIF V4

Novelty OIF V5

Novelty OIF V6

Novelty OIF V7

Novelty OIF V8

Novelty OIF V9

Novelty OIF V10

Novelty OIF V11

Novelty OIF V12

Novelty OIF V13

Novelty OIF V14

Novelty Aust — Iraq Flags

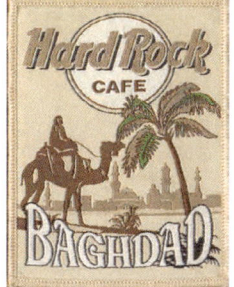
Novelty Hard Rock Cafe Baghdad

Novelty South Park V1

Novelty South Park V2

Novelty South Park V3

Novelty I'm Bored

Novelty Send Me Home

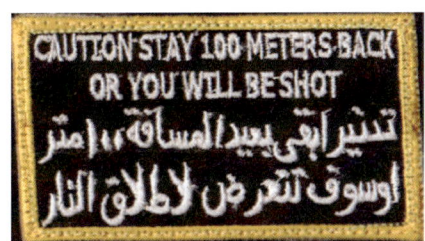
Novelty Warning

Novelty Stay Back 100m Large

Novelty Stay Back 100M Coloured

Novelty Stay Back 100M Subdued

Novelty Ribbon Bar V1

Novelty Ribbon Bar V2

Novelty Ribbon Bar V3

Ali Air Base 407th Squadron Medical Element

IRAQ

Operation Inherent Resolve — Headquarters

Operation Inherent Resolve V1

Operation Inherent Resolve V2

Operation Inherent Resolve V3

Operation Inherent Resolve V4

Operation Inherent Resolve Rubber V5

Novelty Operation Inherent Resolve

Novelty Operation Inherent Resolve

IRAQ

Operation Inherent Resolve — TAJI

TG — TAJI 1 V1

TG — TAJI 1 Rubber V2

TG — TAJI 1 V3

TG — TAJI 1 V4

TG — TAJI 1 V5

TG — TAJI 1 Warhorse rubber

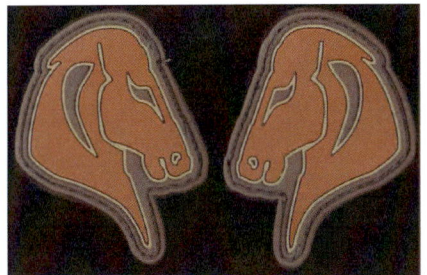
TG — TAJI 1 Opposing Warhorse rubber

TG — TAJI 1 B Sqn 2/14 LHR Colour

TG — TAJI 1 C Sqn 2/14 LHR Colour V1

TG — TAJI 1 C Sqn 2/14 LHR Colour V2

TG — TAJI 1 C Sqn 2/14 LHR Colour V3

TG — TAJI 1 C Sqn 2/14 LHR Subdued

TG — TAJI 1 Training Team 1

TG — TAJI 1 War Horse 4

TG — TAJI 1 ANF Rubber

TG — TAJI 1 NZNF KIWI rubber

TG — TAJI 1 NZNF rubber

TG — TAJI 1 NZ KIWI round rubber V1

TG — TAJI 1 NZ KIWI round rubber V2

TG — TAJI 1 NZ Fern Leaf rubber

TG — TAJI 2 V1

TG — TAJI 2 V2

TG — TAJI 2 V3

TG — TAJI 2 V4

TG — TAJI 2 V5

TG — TAJI 2 V6

TG — TAJI 2 V7

IRAQ

TG — TAJI 2 Coin Design

TG — TAJI 2 Training Task Unit

TG — TAJI 2 Training Team 3

TG — TAJI 2 Training Team 4

TG — TAJI 2 CAV ANTFW

TG — TAJI 2 Q Store Gun Runners

TG — TAJI 2 Training Team 4 Name Tags

TG — TAJI 2 ANF ANZAC Day Iraq 2016 MC

TG — TAJI 2 ANF ANZAC Day Iraq 2016 Tan

TG — TAJI 2 Iraqi Army 15th SOS

TG — TAJI 3 7 RAR BG K-092 Dhobi No

TG — TAJI 3 7 RAR BG UCP

TG — TAJI 3 Training Task Unit

TG — TAJI 3 Training Task Unit Helmet

TG — TAJI 3 K50

TG — TAJI 3 Novelty Riddler Kilo Fun Zero

TG — TAJI 3 IRAQ PARA V1

TG — TAJI 3 IRAQ PARA V2

TG — TAJI 4 1 AR T-127 Dhobi

TG — TAJI 4 1 AR UCP

TG — TAJI 4 Training Task Unit Coloured

TG — TAJI 4 Training Task Unit Tan

TG — TAJI 4 Sallyman

TG — TAJI 5 3 RAR BG K-145 Dhobi

TG — TAJI 5 3 RAR BG UCP

TG — TAJI 5 3 RAR BG V2

TG — TAJI 5 Training Task Unit

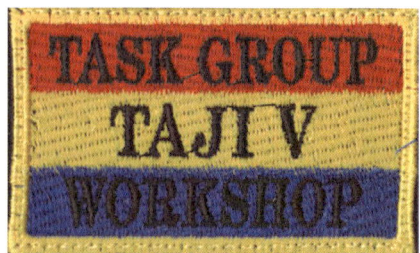
TG — TAJI 5 RAEME Workshops

TG — TAJI 5 EVAC

TG — TAJI 5 ANF V1

IRAQ

TG — TAJI 5 ANF V2

TG — TAJI 5 ANF V3

TG — TAJI 6-12 V1

TG — TAJI 6-12 V2

TG — TAJI 6-12 V3

TG — TAJI 6 ANF

TG — TAJI 6-12 ANF V1

TG — TAJI 6-12 ANF V2

TG — TAJI 6-12 ANF V3

TG — TAJI 6-12 Aust/Iraqi Flag

TG — TAJI 6-12 ANF Subdued V1

TG — TAJI 6-12 ANF Subdued V2

TG — TAJI 9 ANF ANZAC Day Iraq 2019

TG — TAJI 9 NZNF ANZAC Day Iraq 2019

TG — TAJI A SQN 2/14 LHR

TG — TAJI Special Forces VIP Protection Team

TG — TAJI Guardian Angels V1

TG — TAJI Guardian Angels Coloured V2

TG — TAJI Guardian Angels Subdued V3

TG — TAJI Arabic Name Tag

TG — TAJI ANF Punisher Cam

TG — TAJI ANF Punisher Shield Cam

TG — TAJI ANF IR V1

TG — TAJI ANF IR V2

TG — TAJI ANF/NZNF Map of Iraq

TG — TAJI ANF V1

TG — TAJI Aust/Iraqi Flag Coloured V1

TG — TAJI Aust/Iraqi Flag Coloured V2

TG — TAJI Aust/Iraqi Flag Coloured V3

TG — TAJI Rank Aust LCPL

IRAQ

TG — TAJI Rank Aust CPL

TG — TAJI Rank Aust SGT

TG — TAJI Rank Aust WO2

TG — TAJI Rank Aust WO1

TG — TAJI Rank Aust LT

TG — TAJI Rank Aust CAPT

TG — TAJI Rank Aust MAJ

TG — TAJI Rank Aust/Iraqi Major

TG — TAJI Rank Aust LTCOL

TG — TAJI Rank Aust/Iraqi Lt Colonel

TG — TAJI Rank Aust COL

TG — TAJI Rank Aust/Iraqi Colonel

TG — TAJI Rank 4 Bars DPDU

TG — TAJI NZ ATTI

TG — TAJI NZNF KIWI Shield Cam

IRAQ

TG — TAJI NZ KIWI Rect Cam

TG — TAJI NZ WW1 100 years Black

TG — TAJI NZ WW1 100 Years Green V1

TG — TAJI NZ WW1 100 Years Green V2

TG — TAJI NZ Fern & Southern Cross Coloured

TG — TAJI NZ Fern & Southern Cross Cam

TG — TAJI NZ Fern Leaf rubber

TG — TAJI NZ KIWI round Black V1

TG — TAJI NZ KIWI round Black V2

TG — TAJI NZ KIWI round Cam

TG — TAJI NZ KIWI round Grey V1

TG — TAJI NZ KIWI round Grey V2

TG — TAJI NZ KIWI round Tan

TG — TAJI NZNF V1

TG — TAJI NZNF V2

IRAQ

TG — TAJI NZNF V3

TG — TAJI Unknown

TG — TAJI Novelty ANF

TG — TAJI Novelty ANF Captain Australia

TG — TAJI Noverty Combat Cigar Club

TG — TAJI Novelty Never Go Full Retaji V1

TG — TAJI Novelty Never Go Full Retaji Cam V2

TASK GROUP 633.2 Australia/New Zealand

TG — TAJI Iraqi Army ISOF

TG — TAJI Iraqi Army ISOF/2

IRAQ

IRAQ

Operation Inherent Resolve — Novelty

Novelty ISIS Hunting Club

Novelty Make Iraq Great Again

Novelty Make Mosul Great Again

Novelty Enjoy Freedom Baghdad

Novelty Kalashnikova Classic

IRAQ

Operation Operation Okra — SOTG

TG 632 SOTG — IRAQ V1

TG 632 SOTG — IRAQ V2

TG 632 SOTG — IRAQ V3

TG 632 SOTG — IRAQ V4

TG 632 SOTG — IRAQ V5

TG 632 SOTG — IRAQ V6

TG 632 SOTG — IRAQ Rot 1 B Coy 2 Cdo MC V1

TG 632 SOTG — IRAQ Rot 1 B Coy 2 Cdo MC V2

TG 632 SOTG — IRAQ Rot 2 A Coy 2 Cdo Shield

TG 632 SOTG — IRAQ C Coy 2 Cdo MC ANF

TG 632 SOTG — IRAQ C Coy 2 Cdo

TG 632 SOTG — IRAQ D Coy 2 Cdo ANF V1

TG 632 SOTG — IRAQ D Coy 2 Cdo ANF V2

TG 632 SOTG — IRAQ D Coy 2 Cdo Rubber

TG 632 SOTG — IRAQ D Coy 2 Cdo Shield

TG 632 SOTG — IRAQ Call Sign 6CA

TG 632 SOTG — IRAQ Call Sign 82B

TG 632 SOTG — IRAQ Call Sign UA

SOTG Aust/Iraq Flag Coloured V1

SOTG Aust/Iraq Flag Coloured V2

SOTG Aust/Iraq Flag Subdued V1

SOTG Aust/Iraq Flag Subdued V2

MEAO

Headquarters

ANHQ Round V1

ANHQ Round V2

ANHQ Round V3

JTF 633 Coloured Round V1

JTF 633 Coloured Arabic Round V2

JTF 633 Subdued Round V1

JTF 633 Subdued Round V2

JTF 633 Subdued Round V3

JTF 633 Subdued Round V4

JTF 633 Subdued Round V5

JTF 633 Coloured V1

JTF 633 Coloured V2

JTF 633 Coloured V3

JTF 633 Coloured V4

JTF 633 Coloured Arabic V5

JTF 633 Coloured V6

JTF 633 Coloured V7

JTF 633 Subdued V1

JTF 633 Subdued V2

JTF 633 Subdued V3

JTF 633 Subdued V4

JTF 633 Subdued V5

JTF 633 Subdued Arabic V6

JTF 633 Khaki V1

JTF 633 Khaki V2

JTF 633 ADFPP

JTF 633 Tri Colour V1

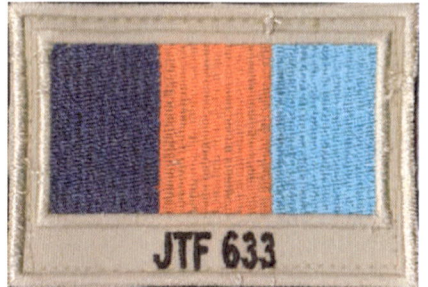
JTF 633 Tri Colour V2

JTF 633 Tri Colour V3

JTF 633 Tri Colour V4

JTF 633 Tri Colour FAKE

JTF 633 Tri Colour V5

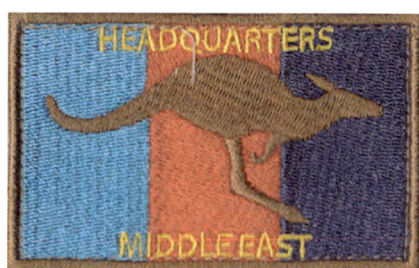
JTF 633 Tri Colour V6

JTF 633 Tri Colour V7

JTF 633 Operation Accordion V1

JTF 633 Operation Accordion V2

MEAO

HQ JTF 633 Watchkeepers Homer V1

HQ JTF 633 Watchkeepers Homer V2

HQ JTF 633 Watchkeepers Skull & Crossed Bones V1

HQ JTF 633 Watchkeepers Skull & Crossed Bones V2

HQ JTF 633 C-IED

JTF 633 CAUS-2 16 ALR

JTF 633 Novelty

MEAO

Engineers

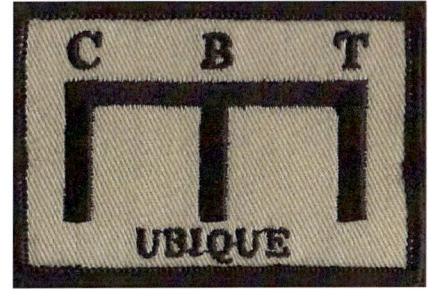
JTF 633 Engineers CBT V1

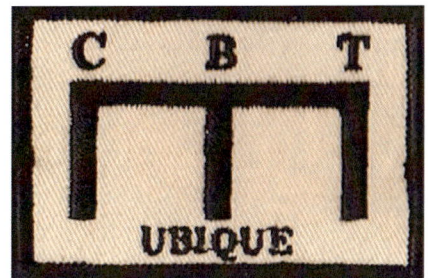
JTF 633 Engineers CBT V2

JTF 633 Engineers CMT V1

JTF 633 Engineers CMT V2

JTF 633 Engineers GTM

JTF 633 Engineers PMT V1

JTF 633 Engineers PMT V2

JTF 633 Engineers PMT V3

JTF 633 Engineers PRT

JTF 633 Engineer Support Element V1

JTF 633 Engineer Support Element V2

JTF 633.16 Engineers FEU

MEAO

FLLA FSU FSE

Force Level Logistics Asset Kuwait

Force Level Logistics Asset Coloured V1

Force Level Logistics Asset Coloured V2

Force Level Logistics Asset Coloured V3

Force Level Logistics Asset Subdued V1

Force Level Logistics Asset Subdued V2

Force Level Logistics Asset Subdued V3

Force Level Logistics Asset 1 V1

Force Level Logistics Asset 1 V2

Force Level Logistics Asset 2 V1

Force Level Logistics Asset 2 V2

Force Level Logistics Asset 2 V3

Force Level Logistics Asset 2 V4

Force Level Logistics Asset 2 Subdued

Force Level Logistics Asset 2 Kuwait

Force Level Logistics Asset 3 V1

Force Level Logistics Asset 3 V2

Force Level Logistics Asset 4 V1

Force Level Logistics Asset 4 V2

FLLA 4 Ammo Section

Force Level Logistics Asset 5 V1

Force Level Logistics Asset 5 V2

Force Level Logistics Asset

FLLA Q Store

FIT/FET 2007 V1

FIT/FET 2007 V2

FLLA Iraq

FLLA Novelty Boxing Kangaroo

FLLA Novelty Fun Police V1

FLLA Novelty Fun Police V2

Force Support Unit V1

Force Support Unit V2

Force Support Unit V3

MEAO

Force Support Unit V4

Force Support Unit MC V1

Force Support Unit MC V2

FSU 1 Tpt Coy 1 CSSB

FSU 1 Memento

FSU 1 Pennant

FET 2009

FIEG ANF 2009

FSU 2 Force Extraction Team

FIAG 2

FSU Australian Forces Post Office V1

FSU Australian Forces Post Office V2

FSU EOD

FSU Termites 2009-10

FSU Psych Round

MEAO

FSU Psych

FSU Psych Worry Meter

FSU Novelty

Force Support Element 1

Force Support Element DPCU V1

Force Support Element DPCU V2

Force Support Element MC

Force Support Element V1

Force Support Element V2

Force Support Element V3

Force Support Element V4

Force Support Element AMCU

FSE 2 ANZAC DAY 2015 DPCU

FSE 2 ANZAC DAY 2015 MC V1

FSE 2 ANZAC DAY 2015 MC V2

MEAO

FSE 2 ANZAC DAY 2015 Poppy V1

FSE 2 ANZAC DAY 2015 Poppy V2

MER Support Unit Rot 1

MER Support Unit Coloured V1

MER Support Unit Coloured V2

MER Support Unit Subdued

MER Support Unit — Logistics Element

MER JO5 Finance 2018-19

MER Force Extraction Team 2019

MER Pysch

MER PWS Training Team

MER Deployment Support Team 2021

MER State of Origin maroon Supporter 2021

MER State of Origin Blues Supporter 2021

1 JPAU

MEAO
FCU TCU

TG 633.14 Force Communication Unit DPDU V1

TG 633.14 Force Communication Unit DPDU V2

TG 633.14 Force Communication Unit

FCU TG 633.14 BODY ARMOUR NAME TAG

TG 633.14 Force Communication Unit DPCU

TG 633.14 Force Communication Unit DPCU Subdued V1

TG 633.14 Force Communication Unit DPCU Subdued V2

TG 633.14 Force Communication Unit DPCU Subdued V3

TG 633.14 Force Communication Unit MC Subdued V1

TG 633.14 Force Communication Unit MC Subdued V2

TG 633.14 Force Communication Unit

TG 633.14 Force Communication Unit 1 Round

TG 633.14 Force Communication Unit 2

TG 633.14 Force Communication Unit 6 Round Large

TG 633.14 Force Communication Unit 6 Round Small

TG 633.14 Force Communication Unit 6 Coloured

TG 633.14 Force Communication Unit 6 Subdued

TG 633.14 Force Communication Unit 6

FCU CIS Install Team

FCU 127 Sig Tp V1

FCU 127 Sig Tp V2

FCU 127 Sig Tp (DPDU) V3

FCU Project Survey Team V1

FCU Project Survey Team V2

TG 633.14 ECT DPDU

TG 633.14 ECT 4

FCU 110 Sig Tp

MEAO

FCU 547 Sig Tp EW

FCU HHC Kuwait

TG 633.14 Heron CIS

Theatre Communication Unit Coloured V1

Theatre Communication Unit Coloured V2

Theatre Communication Unit Subdued

Theatre Communication Unit (AMCU) Subdued V1

Theatre Communication Unit (AMCU) Subdued V2

MER CIT-5

Unknown

MEAO

MEAO

Misc

Joint Operations Command (DPDU)

Defence Security Operations Centre CSAAT 2

DLISST V1

DLISST V2

Maintenance Advisory Service V1

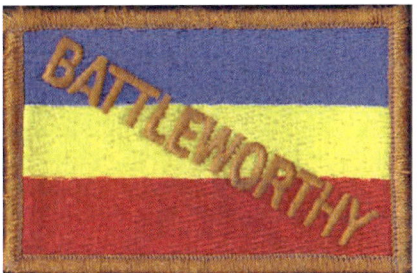
Maintenance Advisory Service V2

Maintenance Advisory Service V3 DPCU

Maintenance Advisory Service V3 DPDU

Maintenance Advisory Service V4 DPCU

Maintenance Advisory Service V4 DPDU

JOC Army Compliance & Assurance Unit MER Audit

Hazard Assesment Team

MEAO Tour De Force 1 V1

MEAO Tour De Force 1 V2

MEAO Tour De Force 2 V1

MEAO Tour De Force 2 V2

MEAO Tour De Force 3 V1

MEAO Tour De Force 3 V2

MEAO Tour De Force 4 V1

Forces Entertainment FACE 6

Forces Entertainment FACE 8

Forces Entertainment FACE

Forces Entertainment V1

Forces Entertainment V2

Forces Entertainment V3

Forces Entertainment V4

Forces Entertainment V5

Forces Entertainment Major League Guitarist

AWM Curator

AWM Combat Curator

AWM Offical Curator

AWM Offical Artist V1

AWM Offical Artist V2

AWM Offical Photographer V1

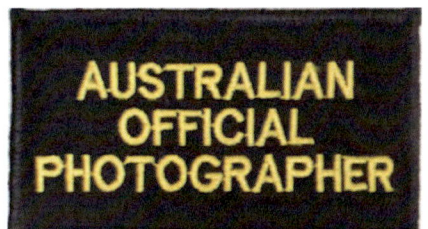
AWM Offical Photographer V2

AWM Offical Photographer V3

AWM Offical Photographer V4

AWM Official Cinematographer

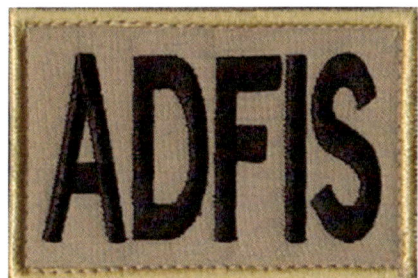
TQ MER ADFIS

TQ Ammunition Technical Officer DPDU Coloured

TQ Ammunition Technical Officer DPDU Subdued V1

TQ Ammunition Technical Officer DPDU Subdued V2

MEAO

TQ Articifer DPCU V1

TQ Articifer DPDU V1

TQ Articifer DPDU V2

TQ Articifer DPDU V3

TQ Dog Handler DPDU

TQ Emergency Response DPDU

TQ Explosive Ordnance Disposal DPDU Coloured V1

TQ Explosive Ordnance Disposal DPDU Coloured V2

TQ Explosive Ordnance Disposal DPDU Coloured V3

TQ Explosive Ordnance Disposal DPDU Subdued V1

TQ Explosive Ordnance Disposal DPDU Subdued V2

TQ Tri Service Explosive Ordnance Disposal DPDU

TQ Tri Service Explosive Ordnance Disposal V2

TQ Tri Service Explosive Ordnance Disposal TAN

TQ Tri Service Explosive Ordnance Disposal V1

TQ Movements V1

TQ Movements V2

JMCC-MEAO Movements V1

JMCC-MEAO Movements V2

JMCC-MEAO Movements V3

Joint Movements Control Office MER ARMY

Joint Movements Control Office MER RAAF

TQ Operator Petroleum DPDU

RAE ACB

Foreign Military Observer

Governor General Foreign Rank 4 Star Equivalent

General Foreign Rank 3 Star Equivalent

Lieutenant General Foreign Rank 2 Star Equivalent

Major General Foreign Rank 1 Star Equivalent

MEAO

ANF

ANF DPDU Brassard Coloured

ANF DPDU Coloured V1

ANF DPDU Coloured V2

ANF DPDU Coloured V3

ANF DPDU Coloured V4

ANF DPDU Coloured Arabic V1

ANF DPDU Coloured Arabic V2

ANF DPDU Coloured Arabic V3

ANF Tan Coloured V1

ANF Tan Coloured V2

ANF Tan Coloured V3

ANF Tan Coloured Arabic V1

ANF Tan Coloured Arabic V2

ANF DPDU Subdued V1

ANF DPDU Subdued V2

ANF DPDU Subdued V3

ANF DPDU Subdued V4

ANF DPDU Subdued V5

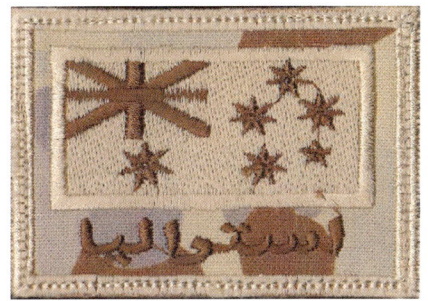
ANF DPDU Subdued Arabic V1

ANF Tan V1

ANF Tan V2

ANF Tan V3

ANF Tan V4

ANF Tan V5

ANF Tan V6

ANF Tan Subdued Arabic V1

ANF Tan Subdued Arabic V2

ANF Tan Subdued Arabic V3

ANF Tan Subdued Arabic V4

ANF Tan Subdued Arabic V5

ANF Tan Subdued Arabic V6

ANF Tan Subdued Arabic V7

ANF Tan Subdued Arabic V8

ANF Lime Subdued Arabic V1

ANF MC Subdued V1

ANF MC Subdued V2

ANF DPCU Coloured V1

ANF AMCU Coloured

ANF AMCU Subdued

ANF SF Subdued V1

ANF SF Subdued V2

ANF SF Subdued V3

MEAO

ANF SF Subdued V4

ANF IR Large V1

ANF IR Small V1

ANF IR Large V2

ANF IR Small V2

ANF IR V3

ANF IR V4

Biscuit — Army V1

Biscuit — Army V2

RAAF GPU ANF

RAAF GPU Ensign

Biscuit RAAF V1

Biscuit RAAF V2

RAN DPDU Ensign

Tab RAAF · Biscuit RAN V1 · Biscuit RAN V2

Biscuit RAN V3 · Tab RAN · Tab AUSTRALIA

MEAO

Novelty

Novelty Team Australia V1

Novelty Team Australia V2

Novelty Team Australia V3

Novelty Team Australia V4

Novelty ANF Team Australia V1

Novelty ANF Team Australia PVC V2

Novelty ANF Team Australia V3

Novelty RAEME Brotherhood on Tour

Novelty Ali Al Salem AB

Novelty Ali Al Salem AB

Novelty Coalition Fire & Rescue

Novelty BRUVCORP V1

Novelty BRUVCORP V2

Raghead Smashfest 2005

Novelty Head Shot

Novelty Infidel Scroll

Novelty South Park Kenny

Novelty Transformers

Novelty Spartans

Novelty Pineapple V1

Novelty Pineapple V2

Novelty Homer To Alcohol !

Novelty Nice Cup of Shut the Fuck Up

Novelty Starbucks I Love Guns & Bacon PVC

Novelty Guns and Coffee

These Colours Don't Run

Unknown 1

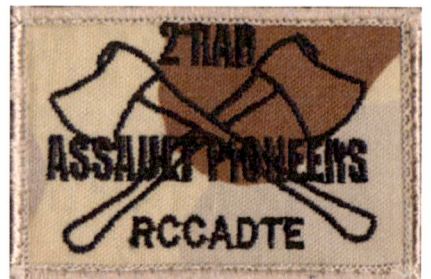
Unknown 2 RAR Assault Pioneers

Unknown 3 Phantom Desert Medics

Unknown 4 DET B

Unknown 5 Combat Engineers C/S 24A

Unknown 6 ACME Removals DPDU

MEAO

Special Forces

TF 12 V1

TF 12 V2

TF 12 V3

TF 12 ANF V1

TF 12 ANF V2

MEAO

RAAF

TE 630.1.8 Roto 1 Coloured AC

TE 630.1.8 Roto 1 Subdued AC

TE 630.1.8 Roto 1 Coloured GC

TE 630.1.8 Roto 1 Subdued GC

TE 630.1.4 Roto 26 3 CRU

TG 633.2 Redback Round

TG 633.2 Redback V1

TG 633.2 Redback V2

TU633.2 Air Group Combat Support Element V1

Hercules Loadmaster

TU633.2 Air Group Combat Support Element V2

TU633.2 Roto 2 V1

TU633.2 V1

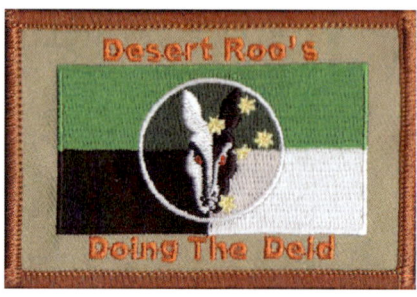
TU633.2 V2

TG 633.2 Australian Air Component V1

TG 633.2 Australian Air Component V2

TG 633.2 Australian Air Component V3

TG 633.2 Australian Air Component V4

MEAO

TG 633.2 Australian Air Component V5

TG 633.2 Australian Air Component V6

TG 633.2 Australian Air Component V7

TG 633.2 MER Combat Support Unit V1

TG 633.2 MER Combat Support Unit V2

TG 633.2 MER Combat Support Unit V3

TU633.2.2 MER Combat Support Unit 4 Expeditionary Health Facility

CSU TG 633.2 What Bag

TG 633.2 MER Air Component Command Element Round

TG 633.2 MER Air Component Command Element Rect

MER Expeditionary Airbase Operations Unit V1

MER Expeditionary Airbase Operations Unit V2

MER Expeditionary Airbase Operations Unit Roto 2

TG 633.2 MER Air Mobility task Group Round

TG 633.2 MER Air Mobility task Group V1

TG 633.2 MER Air Mobility task Group V2

TU 633.2.3 AC

TU 633.2.3 GC

TG 633,4 Australian Air Group AC

TG 633,4 Australian Air Group GC V1

TG 633,4 Australian Air Group GC V2

TG 633,4 Australian Air Group GC V3

TU 633.2.6 C-17 Round AC

TU 633.2.6 C-17 GC V1

TU 633.2.6 C-17 GC V2

Op Okra Air Task Group GC 630 AC V1

Op Okra Air Task Group 630 GC V1

Op Okra Air Task Group 630 AC V2

Op Okra Air Task Group 630 GC V2

Op Okra Air Task Group 630 GC V3

MEAO

Op Okra Air Task Group 630 AC V4

Op Okra Air Task Group 630 GC V4

Op Okra AEW&C

Op Okra Maintainers Roto 10

1 Sqn Round

2 Sqn — Crew 2

2 Sqn — Crew 6

2 Sqn — Round

3 Sqn — Oval V1

3 Sqn — Oval V2

3 Sqn — Roto 8

11 Sqn CRU 6

33 Sqn — Round V1

33 Sqn — Round V2

35 Sqn — Round V1

36 Sqn — Round V1

36 Sqn — Round V2

36 Sqn — Round V3

36 Sqn — Round V4

36 Sqn — Round V5

36 Sqn AC V1

36 Sqn AC V2

36 Sqn GC

36 Sqn Crest

37 Sqn — Round V1

37 Sqn — Round V2

75 Sqn — Round Coloured V1

75 Sqn — Round DPDU V2

75 Sqn — Round Subdued V3

75 Sqn — Round Subdued V4

MEAO

OP Okra 77 Sqn

77 Sqn — Round

77 Sqn — Oval

817 Sqn — Round

RAAF Airfield Engineers

Aircraft Refuelers

C-130J-30 Co-Pilot

C130J Electrical & Mechanical

C-130J Novelty

Ali Al Salem AB Kuwait

TG 633.2 Novelty Camel

TG 633.2 Novelty Survivor V1

TG 633.2 Novelty Survivor V2

TG 633.2 Novelty V1

TG 633.2 Novelty V2

TG 633.2 Novelty V3

F-18

Globemaster C-17

Globemaster Aircrew

Globemaster Loadmaster

Hercules C-130

Hercules Aircrew V1

Hercules Aircrew V2

Hercules Instructor V1

Hercules Instructor V2

TQ RAAF Bomb Disposal DPDU V1

TQ RAAF Bomb Disposal V2

TQ RAAF Bomb Disposal V3

TQ RAAF EOD (MC)

RAAF Medic

MEAO

MER RAAF 100 Years

RAAF ANF Subdued

Unknown RAAF Air Base Protection

Unknown RAAF Rot 9

MEAO

RAN

Commander Task Force 158 V1

Commander Task Force 158 V2

RAN Logistic Support Element V1

RAN Logistic Support Element V2

RAN Logistic Support Element V3

HMAS ANZAC Persian SMETS

HMAS ANZAC CIS Dept

HMAS ANZAC Early Warning Missile Defence

HMAS ANZAC Port Watch

HMAS ANZAC Flight 3

HMAS Arunta Flight Crew

FFG-04 HMAS Darwin OPS Crew

HMAS Darwin Flight Crew

HMAS Kanimbla Flight Crew

HMAS Melbourne 2010

HMAS Melbourne Clearance Diver Team 4 EOD V1

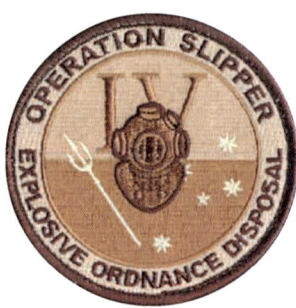
HMAS Melbourne Clearance Diver Team 4 EOD V2

HMAS Melbourne Clearance Diver Team 4 EOD V3

HMAS Newcastle Flight 1

HMAS Parramatta OPS Crew 2006

HMAS Parramatta

MEAO

HMAS Stuart GDP Crew

HMAS Stuart Flight 5

HMAS Warramunga Rot 14

Biscuit Clearance Diver

ANF Navy Clearance Diver

Navy Clearance Diver V1

Navy Clearance Diver V2

Navy Clearance Diver V3

Op Manitou HMAS Ballarat

Op Manitou Rot 3 Flight 3

OTHER OPERATIONS

Nepal

Op Nepal Assist 2015 Army V1

Op Nepal Assist 2015 Army V2

Op Nepal Assist 2015 RAAF V1

Op Nepal Assist 2015 RAAF V2

OTHER OPERATIONS

Pakistan

JTF 632 Op Pakistan Assist Coloured

JTF 632 Op Pakistan Assist Subdued

JTF 636 Op Pakistan Assist II

OTHER OPERATIONS

Sinai

Multinational Force Sinai V1

Multinational Force Sinai V2

Multinational Force Sinai V3

Multinational Force Sinai V4

Multinational Force Sinai V5

Multinational Force Sinai V6

Multinational Force Sinai V7

Multinational Force Sinai V8

Multinational Force Sinai V9

Multinational Force Sinai V10

Multinational Force Sinai Fantasy

Multinational Force Sinai Beret Badge

AUSCON MFO

Novelty MFO Sinai

OTHER OPERATIONS

Ukraine Support

Operation Kudu Coloured

Operation Kudu Subdued

Operation Kudu — Australia United with Ukraine V1

Operation Kudu — Australia United with Ukraine V2

Operation Kudu — Australia Supporting Ukraine V1

Operation Kudu — Australia Supporting Ukraine V2

Other Operations

UNITED NATIONS

United Nations V1

United Nations V2

United Nations V3

United Nations Rect V1

United Nations Rect V2

United Nations Rect V3

United Nations Rect Unknown

United Nations Beret Badge V1

United Nations Beret Badge V2

United Nations Officers Beret Badge V1

United Nations Officers Beret Badge V2

United Nations Officers Beret Badge V3

United Nations Epaulette

United Nations WHOAUST

African Union

Op Hedgerow - Darfur

Sudan UNMISS ASC 2

Sudan UNMISS ASC 3

Sudan UNMIS ASC 7

Sudan UNMIS ASC 12

Sudan UNMIS ASC 13

Sudan UNMIS ASC 14

Sudan UNMIS ASC 15

UNMISS South Sudan Japan Command

Op Aslan UNMISS South Sudan

Op Aslan UNMISS South Sudan

UNMISS South Sudan Team Zulu

United Nations

UNMISS South Sudan ADF ANF

United Nations Korea

United Nations Korea Op Linesmen

United Nations

Glossary

AATTI	Australian Army Training Team – Iraq	KAF	Kandahar Airfield
AC	Aircrew	MAS	Maintenance Advisory Service
ACAU	Australian Compliance & Assurance Unit	MC	MultiCam
		MER	Middle East Region
AMTG	ALMUTHANNA TASK GROUP	MRTF	Mentoring Reconstruction Task Force
ANF	Australian National Flag	MTF	Mentoring Task Force
APEC	Asia-Pacific Economic Cooperation	NATO	North Atlantic Treaty Organization
BG	Battle Group	OBG(W)	Overwatch Battle Group (West)
BN	Battalion	OMLT	Operational Mentoring & Liaison Team
Bty	Battery	Pl	Platoon
Cdo	Commando	RAAF	Royal Australian Air Force
CER	Combat Engineer Regiment	RAAMC	Royal Australian Army Medical Corps
Coy	Company	RAE	Royal Australian Engineers
CT	Combat Team	RAEME	Royal Australian Electrical & Mechanical Engineers
Det	Detachment		
DLISST	Deployed Logistics Information Systems Support Team	RAN	Royal Australian Navy
		RAN	Royal Australian Navy
DPCU	Disruptive Pattern Camouflage Uniform	RAR	Royal Australian Regiment
		Rot/ROTO	Rotation
DPDU	Disruptive Pattern Desert Uniform	RTF	Reconstruction Task Force
Embed	ADF member attached to a Foreign Unit	SASR	Special Air Service Regiment
		SECDET	Security Detachment Iraq
EOD	Explosive Ordnance Disposal	SOER	Special Operations Engineer Regiment
EW	Electronic Warfare	SOTG	Special Operations Task Group
FCU	Force Communication Unit	Sqn	Squadron
Fd	Field	TF	Task Force
FSU	Force Support Unit	TK	Tarin Kowt Afghanistan
GC	Ground Crew	Tp	Troop
GG	Governor General	TQ	Trade/Qualification
GP	Group	UAV	Unmanned Aerial Vehicle
IR	Infrared	UNMISET	United Nations Mission of Support in East Timor
ISAF	International Security Assistance Force		
JTAC	Joint Tactical Air Controller	UNMSS	United Nations Mission South Sudan
JTF	Joint Task Force	V	Variation
JTU	Joint Task Unit		

Bibliography

Army Dress Manual (Chapter 3 Para 3.165 to 3.167)

https://www.army.gov.au/sites/default/files/2023-08/Army-Dress-Manual-AL5.pdf

www.ingramcontent.com/pod-product-compliance
Lightning Source LLC
Chambersburg PA
CBRC092341290426
44110CB00008B/182